The Congress of Phoenix

Rethinking Atlantic Security and Economics

Edited by Gerald Frost

The AEI Press

Publisher for the American Enterprise Institute
WASHINGTON, D.C.
1998

Available in the United States from the AEI Press, c/o Publisher Resources Inc., 1224 Heil Quaker Blvd., P.O. Box 7001, La Vergne, TN 37086-7001; call toll free 1-800-269-6267. Distributed outside the United States by arrangement with Eurospan, 3 Henrietta Street, London WC2E 8LU England.

ISBN 0-8447-4052-7

1 3 5 7 9 10 8 6 4 2

The AEI Press
Publisher for the American Enterprise Institute
1150 17th Street, N.W., Washington, D.C. 20036

Printed in the United States of America

Contents

INTRODUCTION, *Gerald Frost* 1

PART ONE
KEYNOTE SPEECHES

IS THE ATLANTIC COMMUNITY OBSOLETE? *Paul Wolfowitz* 7

DEFENDING THE TRANSATLANTIC WORLD, *Jon L. Kyl* 20

SHAPING THE WEST'S DEFENSES FOR THE TWENTY-FIRST CENTURY, *Margaret Thatcher* 29

PART TWO
COMMENTARY

THE NEW ATLANTIC INITIATIVE, *Christopher DeMuth* 41

THE NATURE OF ATLANTICISM, *Christopher DeMuth, Jon L. Kyl, Charles Powell, Max M. Kampelman, Michael Zantovsky, Brian Beedham, and John O'Sullivan* 42

U.S.-EUROPEAN TIES, *Thomas J. Duesterberg* 49

EUROPEAN POLITICAL INTEGRATION, *Charles Powell, Richard Burt, Dan Quayle, and Margarita Mathiopoulos* 51

A CHANGING BALANCE OF POWER, *Max M. Kampelman, Josef Joffe, and Margarita Mathiopoulos* 53

NATO REFORM, *Christopher Cox, William E. Odom, Rainer Schuwirth, and Richard N. Perle* 55

MEANS AND RATIONALE FOR NATO ENLARGEMENT, *Mark P. Lagon, Jeremy D. Rosner, Mira Baratta, Ojars Kalnins, Douglas J. Feith, Bruce P. Jackson, Valois V. Pavlovskis, Charles Gati, Maciej Kozlowski, Josef Joffe, Brian Beedham, William E. Odom, and Dan Quayle* 58

COSTS OF NATO ENLARGEMENT, *Jeremy D. Rosner, Robert Mroziewicz, Réka A. Szemerkényi, K. Eric Tygesen, Kalev Stoicescu, and Maciej Kozlowski* 67

THE NATO-RUSSIA FOUNDING ACT, *Christopher Cox, Michael Zantovksy, Peter W. Rodman, Josef Joffe, Richard N. Perle, Brian Beedham, Rainer Schuwirth, Frank J. Gaffney, Jr., and Charles Gati* 73

RUSSIA, *Christopher Cox, Max M. Kampelman, Radek Sikorski, Josef Joffe, Brian Beedham, William E. Odom, Dorel Sandor, Martin Sieff, and Dan Quayle* 79

UKRAINE AND THE BALTICS, *Christopher Cox, Anton D. Buteyko, Ojars Kalnins, Martin Sieff, and Dan Quayle* 86

BOSNIA, *Max M. Kampelman and Richard N. Perle* 88

TURKEY, *Richard Burt, Norman Stone, and Margarita Mathiopoulos* 89

MISSILE PROLIFERATION AND STRATEGIC DEFENSE, *Martin Sieff and Dan Quayle* 91

CHINA, *Dan Quayle* 93

APPENDIX A: AGENDA FOR THE CONGRESS OF PHOENIX 95

APPENDIX B: ATTENDEES 99

EDITOR AND CONTRIBUTORS 103

Introduction

Gerald Frost

A word of historical background may be helpful to those not familiar with the origins or activities of the New Atlantic Initiative (NAI). This is a nongovernmental network comprising scholars, former and current public servants, business leaders, and journalists reflecting a wide range of political beliefs and allegiances. It came into existence out of a common apprehension that because of the prevailing mood of introspection on both sides of the Atlantic, Europe and America would drift apart—to the profound detriment of both—unless a serious and sustained attempt was made to apply Atlanticist principles to the conditions of the post–cold war era.

A busy program of conferences, meetings, and publications helped establish three fundamental aims: to anchor the new—in some cases still fragile—democracies of Central and Eastern Europe more securely to Western institutions, to ensure that the European Union's political development proceeded in a way that did not damage U.S.-European relations, and to create a transatlantic free trade area.

The pages that follow comprise the edited text of the Congress of Phoenix held under the auspices of the NAI on May 16–18, 1997, with the aim of defining the means by which these objectives might be met. The congress program and a list of participants are contained in the appendixes.

In preparing the text for publication, I have aimed to provide an accessible guide to the views of participants on central themes and topics that will have an appeal extending beyond merely those who were present at the congress. For this reason, the text has been organized by subject rather than chronologically, although the keynote addresses by Margaret Thatcher, Jon L. Kyl, and Paul Wolfowitz have been given in their original form; my apologies are due to contributors if the form and shape of their remarks have suffered in the process or if some original nuances have been lost. They are owed in equal measure to those whose comments could not be accommodated within the present format.

As the title implies, the purpose of the Congress of Phoenix was to stimulate public debate about ways of strengthening the institutions of Atlantic cooperation so that they meet the needs of a greatly altered economic and security environment. As one participant pointed out, the road to hell may well be paved with new intentions: it is therefore important to stress that the "new" in the title of the New Atlantic Initiative should not be seen as implying a desire for novelty; it is the initiative and circumstances of international politics that are new, not Atlanticism or the principles underlying it. The aim of the NAI is to help sustain both.

A year earlier, at the Congress of Prague, the first major public event staged by the NAI, those taking part had expressed a common fear: that while history remained open and therefore susceptible to efforts to change its course in this, or indeed, other directions, this situation would not last indefinitely. As a result, there was a serious danger that through inertia and neglect, a unique opportunity to expand the area of human liberty and promote structures serving the interests of international stability was being lost. Meanwhile, the nations of Eastern and Central Europe remained in a kind of political limbo, unsure of their own futures as well as the intentions and wishes of the political community that they had bravely struggled to join during the cold war years.

But by the time discussion of such matters moved from the glittering chambers of the Cernin Palace to the air-conditioned luxury of the Arizona Biltmore, it was clear that a start had been made in laying the institutional foundations of the post–cold war order. Evidence for this could be found in the detailed and authoritative accounts by representatives of the U.S. and European governments of the plans—to be announced at the Madrid Summit in July—to begin the negotiations to admit the first new applicants to NATO. That aspiration had enjoyed the ardent support of nearly all those involved with the NAI since its beginnings.

Equally welcome were the accounts from representatives of the

Polish, Hungarian, and Czech governments of the steps being taken to meet their expected obligations as new NATO members and their estimates of the likely costs involved.

News that a NATO-Russia Founding Act, copies of which circulated rapidly among participants following its signing just days earlier, would reportedly allow Russia influence over NATO security arrangements, received rather more mixed responses. Those ranged by degree from the assertion that, given Western prudence, the document would provide an improved channel of cooperation and communications with Russia to the view that it was in fact a cold war stratagem: the tortured prose might have been smartened up to meet the new circumstances, but the aim was still to undermine and destroy NATO.

NATO enlargement, the most suitable means to preserve NATO cohesion and clarity of purpose in the wake of a historic victory, and the likely impact of the Founding Act were inevitably the subjects that dominated discussion. But some of the most interesting debate focused on the future political development of Russia, the impact of Asian economic strength on the international order, and the challenge presented by an increasingly powerful China; there were also stimulating debates on the effectiveness of economic sanctions and on institutional constraints to economic growth.

It is not possible to convey the full range of opinions expressed at Phoenix in this brief publication. It will have fulfilled its purpose, however, if it conveys the views of those present on a range of issues whose outcome will shape a still fluid international order at a remarkably interesting stage of history. It will have fulfilled a second important function if it provides a footnote to an attempt to ensure that that future reflects both the best traditions of Atlantic engagement and the values upon which the Atlantic Community is based.

PART ONE
Keynote Speeches

Is the Atlantic Community Obsolete?

Paul Wolfowitz

I have chosen to address the question of whether the Atlantic Community is obsolete. It might be thought that we have already answered that question by coming all the way to Phoenix to discuss how to deepen the Atlantic Community. Nevertheless, it is worth thinking about whether and why the Atlantic Community is still relevant in an era very different from those earlier ones in which the Atlantic Community proudly triumphed, first over fascism and Nazism and later over communism.

I want to address three specific questions and let me begin by summarizing my answers to each of them. First, is the Atlantic Community obsolete, now that foreign policy has become mostly a matter of trade and economics, and the people in most of the advanced democracies think that foreign policy matters little? My answer, in short, is no. Because that conclusion is dangerously premature, and the task of leadership, particularly in the Atlantic Community, is to persuade the public that the stakes are still large.

Second, is the Atlantic Community obsolete, now that power in the world is shifting so dramatically to Asia? Again, my answer is no. But I would add that this shift in power constitutes the greatest foreign policy challenge of the first part of the next century for the whole world, including for the Atlantic Community. Our chances of meeting that challenge successfully will be greatly improved if the Atlantic Community can address Asia in a unified way as a community of shared values, rather than as a disparate collection of competing mercantile states.

Third, and finally, is the Atlantic Community merely an insurance policy against a revived Russian threat which, even at its worst, will never pose the mortal challenge that the old Soviet Union did? My answer to that is a qualified no. But the Atlantic Community may be judged to have failed if it merely contains a potential Russian threat. The real goal must be to create a stake for a new Russia in the security and prosperity of Europe and to make that same Russia a partner in promoting security in Asia. Let me elaborate briefly on each of these points.

Why Foreign Policy Still Matters

Discussing the first question requires making some assumptions about the future, and I do this with some trepidation. The NATO summit in London in July 1990 was convened to declare an end to the cold war and to offer a new relationship to the Soviet Union and other former adversaries. It was a time of considerable euphoria, and Prime Minister Margaret Thatcher opened the summit with these words: "Europe stands today on the dawn of a new era, as promising in its own way as 1919 and 1945."

The irony was unmistakable, and yet I think there were some among her listeners who failed to notice the warning that the promise of the new era could vanish just as it had in those two earlier times. But I doubt that even Prime Minister Thatcher imagined that less than one month later, Iraq would invade Kuwait and the United States would begin the deployment of more than half a million troops overseas to deal with a new threat.

Yogi Berra, the famous American baseball player and coach who had a way of stating home truths in unintentionally humorous ways, is supposed to have said once that "making predictions is very difficult—especially about the future." As a matter of fact, some people have trouble making predictions about the past, like the revisionist historians who are still trying to prove—even now, in the face of mountains of contrary evidence coming out of Kremlin archives—that the United States was responsible for the cold war.

Sometimes, in fact, history seems to be a better guide to the future than our so-called intelligence estimates are. When I worked in the Pentagon for the first time, in the mid-1970s, I conducted a study of security problems in the Persian Gulf. When we asked the intelligence people about the Soviet threat to the region, we were told not to worry about the twenty-three divisions in the southern Soviet Union near the borders of Iran and Afghanistan. They were all very low-readiness units—made up mostly of reservists—and there was no sign of any Soviet military intentions to the south. Yet those were the same low-readiness divisions with which the Soviets invaded Afghanistan, just two years later. Similarly, when we asked about the large Iraqi military superiority over Kuwait and Saudi Arabia, we were told that one Arab country would never invade another.

So we decided to dispense with an intelligence annex for our study and did a historical annex instead. In this we looked at the three previous Soviet invasions of Iran, in 1921, 1941, and 1946, and at the mobilization of the Iraqi army on the Kuwaiti border in 1961, a move that was countered by the rapid deployment of a brigade of Royal Marines. There are times when history can be a better guide than "intelligence."

It is not that history necessarily repeats itself—not even in the sense that Marx once said, writing about the second emperor Napoleon, "the first time as tragedy, the second time as farce." But history can make us think. As a friend of mine once put it, "History has more imagination than all the scenario writers in the Pentagon." Or in the State Department, for that matter.

I would like to take you "back to the future" for a moment, to illuminate our situation today. We might look at the periods immediately following World War I or World War II. Like the present, those periods both displayed a similarly complacent belief that the end of a major conflict meant the end of all major conflicts. But I do not believe that the greatest danger facing us in the coming era is from another ideologically driven threat like fascism or communism. As we think about our own future, it may be more illuminating to look at the end of the nineteenth century, not because of a coincidence of dates but because both periods have two important features in common: both are periods marked by extraordinary economic growth and by the optimism that growth generates, and both are periods marked by the emergence of new major powers.

Perhaps the single most important phenomenon of our time is the creation of new wealth on a scale and at a pace that is without historical precedent. This is the result of a revolution in technology that is seeing the information age transform the industrial age. It is the result of extraordinary mobility not only of goods but now also of capital that makes specialization possible on a truly global scale. And it is the result, most important, of the triumph of market principles, a process that began in Asia during the cold war and was completed by its end; it was a triumph that left market-based development as the only intellectual contender.

The result has been unprecedented productivity gains in many of the world's economies, in some cases exceeding 5 percent per year on a sustained basis. The resulting wealth creation has lifted hundreds of millions of people out of poverty. It has opened vast new opportunities for global commerce. And it has created a new middle class in a number of countries that hitherto had almost none. That, in turn, is at least partly responsible for another remarkable phenomenon of our time—the triumph of democracy in country after country, including some with no previous history of democratic rule.

I do not often quote President Bill Clinton to agree with him. But he was right when he said in his Second Inaugural Address that "time and chance have put us not only on the edge of a new century, but on the edge of a bright new prospect in human affairs." The dawn of the twentieth century was, similarly, a time of exceptional promise. It, too, was a time of rapid economic growth. Labor productivity grew in the United States and in a number of other industrializing countries at four times the rate of growth at the beginning of the century. This was partly the result of a great expansion of world trade. It was also the result of the great innovations of the industrial revolution, although a number of the most important were still to come at the turn of the century, including the telephone, radio, airplanes, and automobiles.

It is not only economic and technological optimism that have marked the ends of both centuries, but optimism about the prospects for peace as well. If anything, the experience of peace among the major powers was longer and deeper at the end of the last century than it is in our time. There had been no wars between the major powers for nearly thirty years, there had been nothing comparable to the four decades of the cold war, and it had been nearly a century—since the end of the Napoleonic Wars—since the world had experienced war on a global or continental scale.

Indeed, the worst experience of warfare in the last eighty-five years of the nineteenth century was our own American Civil War. By the turn of the century, even that bloody tragedy was a fading memory, replaced in the popular mind by the Spanish-American War, a war lasting less than three months, in which the United States acquired a Pacific empire in a "fit of absent-mindedness." When Commodore George Dewey destroyed the Spanish fleet in Manila Bay in one morning, at the cost of only seven American seamen wounded, the whole country felt a thrill of pride. "Greatest Naval Engagement of Modern Times" was the headline in one newspaper. The general euphoria calls to mind the popular reaction to the victory in the Persian Gulf War.

The optimism of both periods is reflected also in writings that expressed the hopeful expectation that economic and social changes were rendering war obsolete. In our time, that hope has been reflected in the popular reaction to such books as Francis Fukuyama's *The End of History*. At the beginning of the twentieth century, Norman Angell's book *The Great Illusion* became so popular that it went from an obscure printing, paid for by the author, to being a worldwide bestseller that sold more than a million copies. The "illusion" of the title was the belief that nations could profit from war. Not only was this no longer true, Angell argued, but he believed his conclusion was supported by the developments of the preceding forty years, which had "set up a financial interdependence of the capitals of the world, so complex [that it makes] New York dependent on London, London upon Paris, Paris upon Berlin, to a greater degree than has ever yet been the case in history." Were some hypothetical German general to attempt to make the Bank of England a prize of war, "there is no considerable institution in Germany that would escape grave damage" and "the influence of the whole finance of Germany would be brought to bear on the German Government to put an end to a situation ruinous to German trade."

One of Angell's followers, President David Starr Jordan, of Stanford University, argued in 1913 that "the Great War of Europe, ever threatening, . . . will never come. . . . The bankers will not find the money for such a fight, the industries will not maintain it, the statesmen cannot. . . . There will be no general war." He made that statement in 1913. Ironically, Angell himself received the Nobel Prize for Peace in 1933, the year the Nazis came to power in Germany.

Unfortunately, while all this remarkable and peaceful progress was taking place at the end of the last century, the world was grappling with—or, more accurately, failing to manage—the emergence of major new powers. Not only was Japan newly powerful in Asia, but Germany, which had not even existed before the end of the nineteenth century, was becoming a dominant force in Europe.

Today, the same spectacular economic growth that is lifting people out of poverty, that is creating great new commercial opportunities, that is forcing the pace of democratic change is also creating major new economic powers and possibly new military ones as well. This is particularly true in Asia. Small countries in Asia, such as Thailand, the Philippines, and Vietnam, are small only by Asian standards. With populations in the range of 60 to 80 million, they are comparable in size to the great powers of Europe. A unified Korea, which may be on the horizon, would be the size of a major European power, and even without unification South Korea alone is in the process of becoming one of the world's larger economies.

Those, of course, are among the smaller Asian countries. Indonesia, the fourth largest country in the world, is larger than Germany, larger than Russia, small only in comparison with China and India. Those two countries are each three to four times the size of the third largest, the United States.

It is difficult to grasp what China's 1.2 billion population means. In China, three provinces are each larger than unified Germany. The emergence of China by itself would present sizable problems; the emergence of China along with a number of other powers, not only in East Asia but probably in South Asia and Russia as well, presents an extremely complicated equation. In the case of China, there is the ominous element of her outsider status. To hark back to the last turning of a century, the obvious and disturbing analogy is to the position of Germany, a country that felt it had been denied its "place in the sun," that believed it had been mistreated by the other powers, and that was determined to achieve its rightful place by nationalistic assertiveness.

There are enormous differences, of course, between late nineteenth-century Germany and late twentieth-century China. China is an old country, recovering its strength, with a much longer history of involvement in the world. Some argue that the history of China's relatively nonaggressive behavior when it was one of the world's dominant powers (during the Sung and Ming dynasties) gives ground for optimism that a newly powerful China will use its strength with moderation.

But not all the differences are reassuring. China's real sense of grievance at mistreatment by the European powers in the last century and by Japan in this one has a much deeper foundation than Germany's. And in one other respect the similarities to nineteenth-century Germany should give us pause: China is going through a transition from two decades of extremely skillful management of its international relationships to a new leadership of uncertain quality. It was just such a transition from the statesmanship of Bismarck to the incompetence of his successors that is considered to be a principal factor leading to the tragedy of World War I.

It is not inevitable that we will mismanage the emergence of new powers in the twenty-first century as was done in the twentieth. But heaven help us if we do. For it is obvious that the twentieth century did not unfold as it appeared to promise. By its midpoint it was already the bloodiest century in history, and a very large fraction of that bloodshed can be attributed ultimately to the failure to deal with the emergence of the new power of Germany in Europe and of Japan in Asia. The first failure led directly to World War I, a historic disaster of previously unsurpassed scale. World War I, in turn, planted the seeds of Nazism in Germany and Bolshevism in Russia. The first led directly to World War II and the Holocaust; the second led to the crimes of Lenin and Stalin and Pol Pot and to four decades of cold war.

The failure to deal effectively with the emerging Japanese power not only produced the extension of World War II in the Pacific but was directly or indirectly responsible for half a century of horrors in China, some inflicted by the Japanese themselves, some by the Communist regime, whose path to power was opened by the chaos caused by Japanese intervention.

The combined effect of all these events was the violent death of tens of millions of people—more than we are able to count. The twentieth century closes on a remarkably peaceful note, but it has been the bloodiest century so far in human history. If we repeat the history of the twentieth century with the weapons of the twenty-first, then the twenty-first century will become the bloodiest in history and human civilization may not survive. So, the stakes are very large indeed, and the task of leadership should be to explain that to the public. If it can be explained that growing deficits will burden our children and grandchildren, surely it can be explained that war among the major powers in the next century will affect our children and grandchildren as well.

Unfortunately, that kind of leadership is almost nowhere to be found among the leaders of the great democracies today. They seem more inclined to pander to the public feeling that most foreign policy problems will not affect them. It is an attitude reminiscent of Neville Chamberlain's infamous remark on the eve of the Munich Pact, when he lamented that his countrymen were preparing bomb shelters because of "a quarrel among faraway peoples about whom we know nothing." Of course, that attitude of Chamberlain's led to a terrible war that could have been prevented, a war that Winston Churchill called the "unnecessary war."

The Atlantic Community and the Pacific

Let me turn next to the fact that world economic power is shifting to the Pacific, and at a pace that is rather rapid by historical standards. That fact, however, does not make the Atlantic Community obsolete. The key here is understanding the difference between the present and the future. As simple as that distinction might appear, people often seem to confuse the two. Now that Marxism and other central-planning ideologies have been largely abandoned, there does not appear to be anything to prevent the 1.2 billion people of China—or the 600 million in the rest of East Asia, or the billion people of South Asia—from achieving Western levels of per capita output. When that happens, and as that happens, the population numbers will tell the story of relative economic strength.

Today, however, the United States and its democratic allies in Europe and Japan still have the overwhelming preponderance of economic and military power. Moreover, so-called Western values have enormous potential appeal throughout the world, including in Asia. It will make a big difference whether China's economy becomes as large as that of the United States by the year 2015, as some predict, or whether that does not happen until 2030, as some less optimistic forecasts of the Chinese economy project. But even thirty years would represent a very rapid change by historical standards.

The key is whether we can use that time well to create a stable order among the new powers of Asia. Our chances will be greatly improved if the Atlantic Community can function as a unified community of values. Although the purpose of cooperation is strategic, the cooperation required is not primarily military. In fact, the ability to apply the combined economic leverage of the Atlantic Community on major issues may be the best way to avoid military confrontations.

Take the question of China's political future. This is more than a question of human rights. It is also a strategic issue of great importance. A China that rules its own people by force is much more likely to try to impose its will on its neighbors. Conversely, a China that is governed by the real interests of its people is much more likely to find those interests served by a peaceful and equal relationship with its neighbors.

This does not mean that we can dictate the course of political development in other countries, particularly in a great country like China. Our ability to influence may often be quite modest. If we are not careful, our influence may even be negative, as would be the case if the United States were to revoke most favored nation trading status for China. Such a move would not only damage American economic interests, it would also severely damage Hong Kong and Taiwan, as well as the growing private sector within China itself, which is the most important force moving China in a more democratic direction.

While the Clinton administration may have played its hand badly with China, however, in the way it attempted to pursue human rights objectives in the first few years, it would be a serious mistake now to drop human rights considerations entirely, as the administration sometimes seems in danger of doing. While change in China will primarily be a product of forces within China itself, we can help to push change in the right direction—if we can develop leverage that is more carefully targeted, if we can coordinate policies better among the Western countries, and if we do not present our objectives as fundamental challenges to the regime.

The greatest threat to such cooperation among Western countries comes from the intense competition for economic advantage among themselves. If China finds that it can pick off the Western countries one by one, awarding contracts to Airbus when it does not like American policy, or using Japan to block pressure from international financial institutions, or playing Western countries off to dictate the terms of World Trade Organization admission, none of us will have much influence. And the failure to find effective ways of influencing China will lead to more public pressure in the United States to confront China in ways that are likely to be both ineffective and provocative.

Avoiding such a course requires a recognition by all the advanced democracies that something more than commercial advantage is at stake. It requires a recognition by Europeans that developments in Asia, remote as they may appear, are matters not merely of regional concern but of global strategic concern as well. It requires Japan to take a leadership profile on issues of political concern, and not to use legitimate concern about exciting historical animosities as an excuse for pursuing exclusively commercial policies. And it requires a willingness by the United States to accept the compromises that are necessary to produce a common political agenda, rather than indulging in political posturing that may be more emotionally satisfying but less effective.

Current developments give little grounds for optimism on this point, however. Indeed, I have just seen a *Washington Post* story headlined "China Rewards France for Human Rights Backing with Aircraft Contract." But if the Atlantic Community is to be a force in shaping its own destiny, it must address this problem. If the United States finds itself increasingly bearing the military burden of preserving Asian balance and drifts into ineffective confrontations with Asian governments over human rights while our European allies stand on the sidelines and reap rich commercial benefits from a seeming indifference to political principles, we will not only lose the potential benefits of cooperation and tempt aggressive behavior by others but we could tear apart the fabric of the Atlantic Community itself.

The Atlantic Community and the New Russia

Finally, let me address the question of the relation of a New Atlantic Community to a New Russia. Here too, the answer has more to do with Asia than the debate over NATO enlargement has yet recognized. I speak as a strong supporter of NATO enlargement, but not as someone who believes that the purpose of enlargement is to deal with some nearly inevitable reemergence of a Russian threat to Europe. Of course, there is no way to be certain what Russia will become, and that is surely a large part of the reason why Czechs and Poles and Hungarians want to join NATO. And it is hard for me to see how one says no to them. But the Russia of the future will not be the Soviet Union of Stalin and Brezhnev, nor even the Russia of the czars.

What Russia does become will depend on what the Russians

themselves eventually decide. And NATO enlargement will not be the major factor in determining that outcome. I would not dispute that NATO enlargement exacerbates some negative trends in Russia—particularly because there has not been enough clarity and honesty about the difference between the real NATO and the NATO of forty years of Soviet propaganda. But whether one believes, with the opponents of enlargement, that this will be the dominant effect on Russia of bringing new members into NATO, or rather believes, as I do, that this negative effect will become less important as the Russians come to see that the stability that enlargement brings to Central Europe does not threaten them, or exclude them, but actually contributes to their own security and improves their relations with the West, the fact is that the question of Russia's future will be decided much more by what happens inside Russia, and particularly by what happens with the Russian economy.

The most positive outcome, and one that we should try hard to promote, is a Russia that sees its interests advanced by concentrating on development and security within its new borders. A Russia, conversely, that thinks its future lies in reestablishing the czarist or Soviet empires in Europe would clash with the West, even if NATO did not take in new members. A Russia that gets mired down intervening in Central Asia will relearn what the French learned in Algeria and what the Russians should have learned from Afghanistan and Chechnya.

But a Russia that tries to develop as a great nation within its new borders will face real security problems, not from NATO but from other directions. We are told that the Russians cannot accept NATO expansion because of an image of NATO that was created by four decades of Soviet propaganda. It is time, to paraphrase the late Senator J. W. Fulbright, to replace that old myth with the new reality of a Russia that has enormous common interests with the Atlantic Community in promoting stability in Asia, particularly in East Asia. A Poland that is secure within the North Atlantic Alliance will pose no threat to Russia. Would that one could be as certain about China. Even with the best will in the world, it will be difficult to avoid problems on the long border between a resource-rich but sparsely populated Russian Far East and a rapidly growing and overpopulated China.

We should welcome Russia as a significant Pacific power into the Asia-Pacific Economic Cooperation (APEC). We should involve Russia in the resolution of issues of Northeast Asian security, such as

Korea. We should work to develop as much cooperation as possible between Russia and Japan and try to find ways to do this even before the issue of Japan's Northern Territories is resolved. And just as we leaned toward China when it appeared that it was threatened by the former Soviet Union, we should be prepared to lean toward Russia if it is threatened along its long border with China. From this perspective, I view the newly agreed consultative arrangements between NATO and Russia not as a threat to the alliance; not as a price to be paid for NATO enlargement; but as an opportunity to develop a security relationship with Russia that could help us all in the future.

Conclusion

Thus, there are large challenges facing the Atlantic Community. I am an optimist by nature and would like to be optimistic about our ability to meet the challenge of a major shift in the world balance of power. There are some grounds for optimism. Certainly, the spectacular growth of the world's economy is something to feel optimistic about. So too is the broad and strong trend toward greater democracy. But there is nothing to feel good about in the present state of leadership in the Atlantic Community.

Perhaps it is hard to have strong leadership in the absence of a crisis. Crisis has a way of bringing out the best in some leaders, as we saw with George Bush in the Persian Gulf crisis. And strong leaders have qualities that democracies sometimes get impatient with when the crisis is over, as the British public demonstrated so dramatically with Winston Churchill at the end of World War II. We had Winston Churchill and Franklin Roosevelt in a crisis. We had Harry Truman in a later crisis. More recently, we had Ronald Reagan and Margaret Thatcher and Vaclav Havel and George Bush in times of crisis.

There is no crisis today. But if we wait until there is one, it may be too late. Of one thing we can be sure. There will be one simple test of success or failure. If, twenty-five years from now, the world experiences another great war, the coming generation will say that we failed. Looking back in 1938 at the turn of the last century, on the eve of a second world war, Winston Churchill wrote:

> The scale on which events have shaped themselves [since 1895], has dwarfed the episodes of the Victorian Era. Its small wars between great nations, its earnest disputes about su-

> perficial issues, the high, keen intellectualism of its personages, the sober, frugal, narrow limitations of their action, belong to a vanished period. The smooth river with its eddies and ripples along which we then sailed, seems inconceivably remote from the cataract down which we have been hurled and the rapids in whose turbulence we are now struggling.

If the smooth river on which we are sailing now carries us safely through the coming reshaping of world power, we will have succeeded in something of major importance. If instead our successors have to look back wistfully at the end of this century the way that Churchill did at the last one, they will be looking from a very terrible vantage point indeed.

Defending the Transatlantic World

Jon L. Kyl

Lady Thatcher presides over an important school of thought about national security in the United Kingdom. I adhere to a conservative U.S. school of thought that is philosophically akin to hers. This kinship of spirit is a powerful force in the world. It is the very lifeblood of the Atlantic Alliance. That Lady Thatcher in Britain and I in the United States can, without coordination, mull the great national security questions facing the community of democratic states and come to such complementary conclusions is a sign of the vitality of our community—proof of the commonality of our interests—indeed, a symbol of the validity of this grand enterprise: the New Atlantic Initiative.

The Community of Civilized Nations

As participants in this Congress of Phoenix well understand, the invitation that will shortly be extended to several candidates for NATO membership is more than an enlargement of a military bloc. It is also a seminal event for those states as they are formally recognized as part of what can be called the community of civilized nations. That term does not ignore these countries' long and luminous cultural histories. Instead, it recognizes their embrace of democratic and free market institutions that distinguish them as nations with governments that truly respect the rights of their people.

Put another way, the new members of an enlarged NATO will, by joining this organization, be affirming a commitment to a set of moral principles: common concepts of due process of law, freedom of conscience, representative government, and protection of private property.

To be sure, there are nations in this world that are not members of NATO but who are, nonetheless, equally committed to these moral and political convictions. Regrettably, however, there are too many other countries whose governments do not subscribe to such principles. I am sure no one attending the Congress of Phoenix is under any illusion about the continuing threat such states pose to the civilized world. And yet, how frequently are we told—including by some in our nations' ruling elites—that the passing of the cold war has brought an end to strategic dangers to international peace and prosperity.

This false notion has a number of familiar manifestations—and several potentially ominous implications. In discussing them, I will highlight the distinction between policy and strategy. Simply put, policy is what governments do. All governments have policies, but not all have strategies. A strategy is a broader conception. It is what gives coherence to policies in different areas, linking them to the past and aiming them at longer-term goals. It is what ties policies down to the core principles that motivate us as a people and a country. Some of these policies have harmonized with sound strategic thinking, and some that have not.

A Failure to Think Strategically

My first point is that our nations tend to give too little thought to strategic interests and how best to safeguard them. To devise effective strategies, however, we must first have a clear idea of what it is we are intent on securing. The essence of the answer is a way of life secured by institutions of government built on moral principles—and no one has done more to stress the moral component of the national interest than Lady Thatcher. As I said at the 1996 Congress of Prague, "Truly strategic thinkers appreciate that what sustains the Western Alliance is not mere *Realpolitik* but the common attachment of our people to moral and political convictions that we harbor in our heart of hearts."

One great historical example of the effectiveness of a strategic plan borne of such a shared attachment occurred in 1947. At that time, General George C. Marshall, who served as chief of staff of the

U.S. Army during World War II and subsequently as secretary of state, delivered a commencement address at Harvard University. It proved to be an intellectual shot heard 'round the world. His remarks, laying out what became known as the Marshall Plan, codified the American commitment to help Europe rebuild itself.

In endorsing it, the Truman administration demonstrated an appreciation for the economic and political ties binding the United States and Europe, as well as an understanding of the historical and moral consequences of not aiding Europe, including Germany. Winston Churchill called the Marshall Plan "the most unselfish act in history."

While the plan grew out of concerns for postwar economic security, it developed into a political and strategic program to stem the spread of communism in Europe. British Foreign Secretary Ernest Bevin described the plan as "the quickest way to break down the Iron Curtain." He was right. Specifically, the Marshall Plan gave rise to NATO, and a resurgent European Community, and a stronger, closer Atlantic Alliance. That was not just policy; that was strategy in action.

Undue Reliance on Peace Processes

A second problem with current thinking is the exaggerated expectation of the contribution that arms control or peace negotiations with tyrants, liars, and terrorists can make to national or collective security. There is, of course, a danger that, in talking about principles and in voicing skepticism about diplomacy with rogues, one will appear moralistic, idealistic, or otherwise ill-equipped for the unsentimental give-and-take of real-world international relations. Our century's history should have taught us, however, that when democratic powers set aside basic moral judgments in their diplomacy, they wind up damaging themselves and their allies.

Western statesmen have negotiated with Hitler, Stalin, Brezhnev, Ho Chi Minh, Saddam Hussein, and Slobodan Milosevic. These statesmen have, time after time, chosen to ignore or downplay the obvious moral failings of such interlocutors. In some cases, our leaders simply failed to grasp their interlocutors' true, evil nature. In other cases, they rationalized efforts to achieve cooperation with such villains, dismissing moral judgments as being the province of priests, not statesmen. But while peace diplomacy with ruthless, dishonest regimes has often produced agreements, it has failed time and again to produce peace.

If anything, such diplomacy has often served to undermine peace and the security of the democratic parties. Chamberlain's ephemeral success at Munich, Roosevelt's at Yalta, the several U.S. arms control accords with the Soviet Union in the 1970s, and Clinton's Dayton Accords all failed to deliver on their promises of durable peace and genuine security. And they all failed for essentially the same reason: the agreements were concluded with scoundrels who used peace diplomacy to advance or secure their own aggressive ends.

The Chemical Weapons Convention

In this connection, I would like to say a few words about one of the most recent examples of this phenomenon: the Chemical Weapons Convention (CWC). This treaty was recently the subject of a heated debate in the U.S. Senate. I was among those—including four former secretaries of defense, James Schlesinger, Donald Rumsfeld, Caspar Weinberger, and Dick Cheney—who were critical of the CWC. We argued that U.S. security, and that of our allies, would not be enhanced by American ratification of an agreement that even many proponents acknowledged would not be effective, could not be verified, and would be extremely difficult to enforce.

Of special concern was the prospect that nations like Iran, Cuba, China, and Russia might exploit in bad faith the convention's requirements for the "fullest possible exchange of technology," including equipment, materials, and know-how directly relevant to the production of chemical weapons and to chemical defensive activities.

In response, pledges were made that the treaty would not be interpreted in such a way as to exacerbate the danger of chemical proliferation. The problem is that this danger is inherent in the convention's language—and in the prospective participation in the CWC regime of rogue states and their friends.

Let me be clear. The opposition expressed in this country to the Chemical Weapons Convention was not, as some would have you believe, animated by isolationist sentiments. It is preposterous to suggest that individuals like the four above-mentioned former secretaries of defense and Jeane Kirkpatrick, Richard Perle, Doug Feith, or Frank Gaffney are isolationists. I know of few people more committed to the principle of American engagement in world affairs, more serious about the importance of U.S. international leadership, and more conscientious about the grave, and growing, danger posed by the proliferation of weapons of mass destruction than these and other leaders of the fight against the CWC.

Where we disagree with the Clinton administration and other proponents of the CWC is about the value of multilateralist approaches that rely not upon the community of civilized nations but upon the bona fides of the larger so-called international community. The Chemical Weapons Convention—and the multilateral mechanism created to implement it—exemplifies the problem of relying on the good faith of rogue states. And I remain concerned that it will produce the sort of false sense of security to which we know the community of civilized nations is all too susceptible.

Failing to Attend to Real and Growing Threats

A third, and related, problem is the failure of our nations to invest in real military capabilities commensurate with the threat. It is one thing to build defenses and then attempt through treaty negotiations to eliminate the threat. It is quite another to forgo defenses in the hope that they have been made unnecessary by treaties—treaties made with the very parties who threaten us. An example of the latter is the failure to defend adequately against ballistic missiles, the delivery system of choice for chemical and other weapons of mass destruction. At present, neither Europe, nor our friends and forces in Asia and the Middle East, nor the United States enjoys protection against ballistic missile attack.

This sorry state of affairs is a function of the direct and indirect effects of shoddy strategic thinking and misplaced confidence in another arms control agreement—the 1972 Anti-Ballistic Missile Treaty. The ABM Treaty was a product of a different age: a bipolar world in which the two superpowers had a virtual monopoly on long-range ballistic missile technology. Under those circumstances, some found the logic persuasive that U.S. and Soviet missile defenses should be constrained in order to achieve reductions in the two nations' respective offensive missile forces.

Today, the authoritative report *Jane's Strategic Weapons Systems* estimates that there are 13,500 ballistic missiles in service in twenty-six countries, with as many as thirty new missiles under development. Eastern, Central, and Western Europe are increasingly facing missiles in the hands of nations that, in some cases, plainly intend harm to the people, property, and interests of the civilized world.

Unfortunately, the "theater" of North America is also facing a growing menace from non-Russian ballistic missile forces. A year and a half ago, China engaged in what one senior Chinese official called nuclear blackmail against the United States. The blackmail took the form of intimations of a Chinese willingness to attack Los Angeles if the United States interfered in Beijing's campaign of coercion against Taiwan. This episode clearly demonstrated the strategic—as well as the security—implications of America's present, absolute vulnerability to ballistic missile attack.

Even if one discounts the threat of a Chinese missile strike, it would be folly to ignore the malign intentions of North Korea. As Lady Thatcher points out in this volume, we know that North Korea is preparing a ballistic missile of sufficient range to put at risk not only all the Japanese home islands but Alaska and Hawaii as well. (I should note that this Taepo Dong 2 missile, like earlier, less formidable North Korean missiles, might well be available for sale to countries in the Middle East and North Africa that may wish to use them to threaten Europe.)

Those of us who believe that the ABM Treaty should be retired are not, as is often charged, living with a cold war mentality. On the contrary, it is precisely because we appreciate that the cold war is over—and bipolarity is dead—that we seek new security approaches for a world in which major threats to our societies can come from many different quarters.

In such a world there is no compelling argument for preserving the ABM Treaty, to say nothing of expanding its scope and limitations, except that it is in place. If the ABM Treaty were not currently on the books, no one would propose such a ban on missile defenses today.

What Needs to Be Done Now

Both the United States and the rest of the community of civilized nations are in urgent need of near-term, effective, and affordable antimissile defenses. Fortunately, an option is at hand that would permit such defenses to be rapidly and affordably brought to bear. The United States has invested nearly $50 billion to date in the Navy's

AEGIS fleet air defense system. As a result, our country has already largely deployed the infrastructure—the platforms, launchers, missiles, and sensors—necessary to provide an early, global defense against ballistic missile attack.

The Clinton administration is proposing a so-called National Missile Defense program that would be incapable of defending parts of the United States (notably, Alaska and Hawaii). It would be unable to deal with threats involving attacks of larger than a handful of incoming missiles and would be of no assistance whatever in defending our troops and allies overseas. For less money than the administration's favored program would cost, the United States could bring to bear a formidable global missile defense capability by means of the Navy's AEGIS system.

Unfortunately, the contribution the AEGIS system could make to meeting the alliance's near-term requirement for missile defenses is being severely constrained by the ABM Treaty. The Clinton administration favors limitations on the speed at which interceptors can fly or on the data antimissile-equipped AEGIS ships could use to target and intercept ballistic missiles. In its efforts to preserve this treaty, the administration is precluding the introduction of an effective sea-based theater missile defense system.

The Challenge Ahead

The community of civilized nations has a common imperative to bring about the swiftest possible deployment of competent sea-based antimissile defenses, systems that should be permitted to take full advantage of existing technology and available sensors and communications capabilities. Through collaboration with allied navies and command authorities, these sorts of capabilities could begin to protect NATO member states and other friendly nations within a few years' time. By adopting an evolutionary approach, the initial sea-based system optimized for theater missile defense can over time afford better and better protection to the American people as well.

Such an effort should be a priority for all of us who love freedom and who understand what it will take to protect it. We should see the pursuit of such a joint effort to acquire effective ballistic missile defenses as a means of responding strategically and realistically to a common threat. It bears recalling that our alliance effort to deploy intermediate nuclear forces (INF) missiles in the early 1980s was commonly criticized as destabilizing and counterproductive. In the end, however, it proved enormously useful—and not simply in providing for the common defense. The demonstration of will that NATO displayed in implementing that deployment decision gave heart to the alliance and helped to demoralize the hostile ideology that NATO was created to oppose.

In the same way, missile defenses can not only reduce the danger posed by ballistic missiles to all of us. Their deployment may also discourage and, yes, even demoralize militarily ambitious regimes with hostile ideologies. I would call this constructive demoralization. At the very least, the civilized world's acquisition of effective antimissile defenses will diminish the value of ballistic missiles. This could be a most effective technique for combating proliferation.

As we develop missile defenses, we should make clear that this is not an anti-Russian initiative. Indeed, the sorts of defenses contemplated in the AEGIS approach would be incapable of preventing a massive, Soviet-style missile attack. They would instead be suited to preventing smaller, less capable forces from attacking the democratic nations and other states, even including Russia.

Let us remember that on June 17, 1992, Presidents George Bush and Boris Yeltsin issued a joint statement that said, in part: "The presidents agreed that their two nations should work together with their allies and other interested states in developing a concept for [a global protection system against ballistic missile attacks] as part of an overall strategy regarding the proliferation of ballistic missiles and weapons of mass destruction."

This language offers a way ahead. We should reaffirm our desire to pursue what might literally be called a modus vivendi with Russia—one based on mutual assured survival rather than mutual assured destruction. As long as we are clear that the civilized world's urgent need to deploy sensible defenses will proceed with or without such an arrangement, past experience (notably with the INF deployment) suggests that the Russians will be prepared to cooperate.

Conclusion

Current discussion about the Atlantic Alliance focuses mainly on the *who*—who will be admitted to NATO, now and later. I want to ensure that the *why* remains prominent in our minds also. We can never remind ourselves too often of why we are an alliance, why we are a community. There is a philosophical—indeed a moral—purpose to the alliance. It is principle, not policy, that is the fabric of the broader community of democratic states. That fabric can be strengthened by common policies, common enterprise. I suggest that missile defense could be one such enterprise.

There is work yet to be done by our alliance and our community. If we conceive of this work strategically, seeing it in its broadest context—historical and philosophical, rather than just political and military—we maximize our chances for popular support and for success. Our alliance remains an instrument for making the world a safer and a better place.

Shaping the West's Defenses for the Twenty-first Century

Margaret Thatcher

Those of us who live in Europe are often told that the economic success of the Sunbelt states is bringing about a fundamental change in America's political outlook and priorities. We are advised that in the future we must take account of an America that is shifting its gaze from the Atlantic to the Pacific—an America in which new elites are emerging to challenge the assumptions of those from the Eastern seaboard who shaped the Atlanticist outlook of the post–Second World War generation.

As a consequence of these changing demographic and economic realities, the United States will grow weary of its Atlantic partner, giving higher priorities to the more enticing relationships that it is forging in Asia and the Pacific Rim, or so the argument runs. European states must therefore revise their security and trade arrangements. If this then has the effect of encouraging the very trends that Europe is said to fear, this is merely the price of responding to changing realities, so it is said.

But let me make it clear: I believe these arguments to be deeply and dangerously flawed, the product of imprudent extrapolation, weak reasoning, and predetermined conclusions. Moreover, I regard the fact that this congress is being held in Phoenix, with magnificent local support, as tangible evidence that my reservations are well grounded. For it provides a clear indication not only of America's enduring commitment to the Atlantic Alliance but also of the contribution that the processes of American economic change can make in underpinning that commitment.

America's Demographic Shift

Indeed, it seems to me that the significance of America's changing demographic profile has been misunderstood. The remarkable shift in resources to Arizona and the other states of the American Southwest is an indication of America's capacity to renew itself, not a demonstration that it is undergoing a profound change of identity. The character of a nation does not change because some of its most enterprising citizens occasionally change their address: pity the nation in which such mobility does not occur. America's commitment to the European democracies, old and new, is a reflection of its character and its values.

The purpose of this congress is to entrench, extend, and renew our Western inheritance of liberty. More specifically, it is charged with finding ways to deepen the institutions of the Western Alliance. "*Congress* of Phoenix" may have an unexpected ring to it because we tend to assume that congresses are events that take place only in the venerable seats of governments.

But where better to pursue our tasks than in a city that symbolizes America's freedom of enterprise and the entrepreneurial spirit that freedom nurtures? Where better to find an atmosphere conducive to the achievement of our task than in a state that has reaped the rewards of high-risk, high-tech endeavors while remaining faithful to America's best traditions of international engagement?

I may be impressed by the fact that Arizona's increasing economic and political significance is serving to strengthen its ties with its traditional allies in Europe, rather than the reverse, but I am scarcely surprised by it. After all, I worked quite well with the favorite son of another Sunbelt state during the climactic final years of the cold war. The Californian ascendancy of the 1980s may have presaged a change in White House dress code, but not a cooling in the Atlantic relationship; far from it. The immensity of Europe's debt to Ronald Reagan in accelerating the decline of Soviet communism and in healing a divided continent has not been acknowledged. Europeans can best repair this omission by playing a more constructive role in helping to redefine the alliance whose preservation was one of his greatest achievements. This means resisting the allure of certain courses of political action that are incompatible with that aim, a subject to which I shall return.

Atlantic Community Values

Happily, the robust and cogent address of Senator Jon L. Kyl to the Congress of Prague in 1996 dispelled any fear I may have had that on this side of the Atlantic a new generation of Republican leaders would dissipate Ronald Reagan's legacy: Kyl's analysis of the problems faced by the alliance and his proposals for overcoming them were compelling.

Those who believe that the alliance has had its day are apt to lay stress on economic factors. But the Atlantic community is not primarily about economics; it is a matter of common values and of shared historical experience. It is the failure to understand this that encourages the doubters to suppose that in the absence of a single common enemy the alliance has outlived its usefulness. But, even in the matter of economics, the doubters have got it wrong. Economic ties between the United States and European states remain of fundamental importance to all parties, and they are expanding. This is true even of those American states that are alleged to be obsessed with the importance of Asia and the Pacific Rim. In 1993 (the last year for which I have been able to find figures), investment in Arizona by European states amounted to 41 percent of total investments, compared with 40 percent by *all* of Asia.

I do not point to these facts in order to doubt the huge importance of Asian economic success. Indeed, a very great deal depends upon our ability to accommodate the Asian states within a liberal trading regime. I point to these facts to show that there are dangers that can result from rash assessments of its significance.

My main aim today is to offer some thoughts about how Atlantic institutions can be strengthened to meet new security needs before the NATO summit of July 1997. Certainly, the security environment has undergone some profound changes. Unless the reform of those institutions with responsibility for security is based on a clear understanding of their significance, this will heighten instability, rather than the reverse.

As we have moved from a relatively stable, bipolar world to a multipolar one, new sources of instability and conflict have emerged. This was inevitable, since—quite suddenly—a major constraint on the spread of conflict was suddenly removed—the fear that a minor conflict might lead to a nuclear conflagration.

Post–Cold War Dangers

In short, the world remains a dangerous and unpredictable place, menaced by more unstable and complex threats than it was a decade ago. One measure of this instability can be found in the fact that more Europeans have been killed in war during the past five years than in the preceding fifty. Another measure can be found in the alarming proliferation of weapons of mass destruction, a subject on which I shall say more. But because the risk of nuclear annihilation has gone, we in the West have lapsed into a dangerous complacency and relaxed our guard. In almost every Western country, defense spending has fallen and is set to fall still further; during the British general election, campaign defense scarcely rated a mention. Yet defense, as Adam Smith famously wrote, is more important than opulence.

To comfort ourselves that we are doing the right thing, we have increasingly placed our trust in international cooperation to safeguard our future. But international bodies have not generally performed well. Indeed, we have learned what perhaps should have been self-evident: that they cannot perform well unless we refrain from utopian aims, give them practical tasks, and provide them with the means and backing to carry them out.

During the years following the Second World War, NATO amply fulfilled its purpose. The consequence was victory in the cold war. No alliance in history has created a comparable organization or imbued it with a sense of permanence. It is the body through which the Western Alliance finds its best expression, one that has indisputably conferred more benefits upon the citizens of its members than any other international organization. But since the dissolution of the Warsaw Pact, some Western statesmen have been undecided about NATO's purpose, its membership, its methods, and its priorities.

How confident can we be that these uncertainties will be dispelled at the Madrid summit, which, according to its secretary general, Senor Javier Solana, "will mark the moment when the processes of transformation and change converge and the new Alliance emerges"? I find the title for that occasion—"Summit for Euro-Atlantic Security and Cooperation"—less than promising. It partly reflects the current attempt to transfer responsibility from European nation-states to a unitary European state and the associated bid to create a separate European defense identity—separate, that is, from the country that brings the greatest military assets to the alliance, but that curiously is absent from the title.

The ambition to create such an identity is driven entirely by politics, and it rests on two illusions. The first is that the European nations will be prepared to cede responsibility to a federal European state in times of acute crises; they will not. The second illusion is that a strong second pillar to the alliance would ensure greater cohesion and military effectiveness. Quite the opposite is the case: were such an entity to be created, it would inevitably produce antagonism and competition. As I argued in Prague in 1996, for America such a step would transform an ally into a rival—or, at the very least, would permanently threaten to do so. We must not allow NATO to be used as a building block in the creation of a federal Europe. The simple question to be asked of all and any reform is, Will it strengthen our common defenses? If that question becomes the test of relevance, then NATO will maintain the clarity of purpose on which success has been built.

These words are intended as a warning, not as a criticism of the changes that are likely to be determined at Madrid. One such change is both profoundly welcome and overdue: NATO enlargement. The Madrid meeting is expected to start accession negotiations with the aim of admitting the Czech Republic, Poland, and Hungary in 1999.[1] The case for including these three states is overwhelming—they should have been included long before now. Continued tardiness would not only be shortsighted, but morally offensive. It would leave a security vacuum in a part of the world that, it has been noted, has a tendency to produce more history than can be consumed locally.

The accession to NATO of these three states will answer two fundamental needs; it will provide an institutional framework to underpin stability, and it will provide NATO with a common, cohesive purpose, without which it may atrophy.

The Russian Response

I know those who will be meeting in Madrid recognize the force of

1. The NATO summit held in Madrid (July 8–9, 1997) agreed to begin negotiations to admit Poland, Hungary, and the Czech Republic, as Lady Thatcher anticipated.

such arguments. I hope that they will also recognize the dangers of seeking to assuage alleged Russian fears about the purposes of enlargement by seeking in some way to qualify or limit the status of the new members—for example, through a commitment not to base foreign troops on their territory. That step would effectively create a two-tier membership. If NATO credibility is not to be seriously eroded, however, there must be only one category of membership: one conferring equal rights and benefits. Above all, an attack on one must continue to be regarded clearly and unambiguously as an attack upon all.

Diplomats, I know, are not given to taking clear, effective decisions. That's what strong-minded politicians are for. Our strategic intentions with respect to the new members must be as clearly signaled as those with respect to the existing membership. A partial or equivocal security guarantee can sometimes look like a green light to a potential aggressor, and with catastrophic consequences. All of this must be borne in mind when the public case for enlargement is made prior to its ratification by the parliaments of NATO member states. Those in Peoria—or Phoenix or London, for that matter—must understand the binding nature of the guarantees being extended to the new members and the risks and obligations that flow from them. Such understanding cannot wait upon the outbreak of conflict.

Of course, we should seek to engage the Russians in a candid, constructive dialogue about all matters of mutual interest, but there is no plausible reason for bowing to the pressures they have recently been exerting. We should select our words carefully when describing the changes we propose—for fear that extreme Russian nationalists misinterpret them for their own cynical ends. But the Russian people have been provided with irrefutable evidence of NATO's purely defensive character for more than fifty years, and during a period when their leaders threatened to bury us. And to an even greater extent than ourselves, Russia has a powerful interest in the stability of Central Europe—to which NATO enlargement will make a powerful contribution.

My conclusion is that no institutional arrangement linking NATO with its former adversary should be entertained if it obscures its purpose or diminishes its central task. There are many other vital matters to be discussed at Madrid—not least the question and the modalities of operating out of area, and the extremely important matter of reaching an understanding on security issues with Ukraine.

But I have two concerns that do not figure on the Madrid agenda: one general, the other specific, urgent, and compelling. The first and general concern relates to the tendency to stress what is sometimes described as NATO's "political dimension." In a flurry of articles and seminar papers, learned experts refer to the need to "de-emphasize" NATO's defensive aspects and to widen its concerns. I disagree. NATO has always been political, in the sense that the mutual security commitments at its core have reflected political agreement. But the vague terminology to which I take exception points to the risk that NATO's defense capability will be diluted or diminished, and its essential character undermined. The danger is that it will become either another talking shop and photo opportunity—despite impressive competition from other international organizations to fill those roles—or, worse still, a subcontractor to the United Nations. Any attempt to make NATO "politically correct" would not merely be absurd; it would destroy that clarity of mission on which its past success has crucially depended.

The use of language by some of those who write about NATO is particularly revealing. I note that the term *defense* is used less and less, having given way to *security*, a somewhat vaguer term that nevertheless has its uses. But according to an article in a recent issue of *NATO Review,* what is meant by *security* will have to be stretched still further. What is required, according to the author, is a "multidimensional strategy encompassing such issues as the impact of socioeconomic change; the phenomenon of the global financial system; and transboundary environmental issues."

Having failed to demolish Western defenses by other means, it seems that our enemies are trying to achieve their objective by obscuring the issues, and by boring us into submission. My more pressing concern arises from the proliferation of weapons of mass destruction and the ballistic missiles to deliver them—and the seeming determination of Western governments to preserve our vulnerability to future missile attack.

According to the Defence Studies Centre at Lancaster University in Great Britain, thirty-five non-NATO countries have ballistic missiles. As it rightly points out, the greatest potential menace is the five "rogue states," some of which have helped one another's missile programs: Iran, Iraq, Libya, Syria, and North Korea.

According to U.S. sources, all of Northeast Asia and Southeast Asia, much of the Pacific, and most of Russia could be threatened by the latest North Korean missiles. Once they are available in the Middle East, all the capitals of Europe will be in target range. According to present trends, a direct threat to American shores is likely to mature early in the next century.

When I warned of these dangers in March 1996, some commentators suggested that I may have exaggerated: the Korean missile program would surely slow as a result of that country's crippling economic difficulties. But on April 14 came the striking evidence to the contrary: the Japanese foreign minister referred to reports that North Korea had now deployed the Rodong 1 missile, which has a range of 625 miles and is therefore capable of striking any target in Japan. This is an ominous development that threatens regional stability and puts at risk U.S. forces in South Korea as well as those based in Japan. Meanwhile, North Korea continues to develop more advanced, longer-range missiles, with a range of 2,500–4,000 miles, capable of striking the West Coast of America. Since I warned of the risk that tens of thousands of people might be killed by an attack that wise preparation might have prevented, three factors have further convinced me that my concerns were not misplaced:

- We have growing evidence of cooperation between the proliferating states: the Rodong missile that North Korea may now have deployed is believed to be financed by Libya and Iran.
- There are reports that Iran has tested components of a missile capable of striking Israel.
- Western attempts to prevent the sale of advanced Russian military technologies and of nuclear reactors to Iran have quite evidently failed; Russia's *Arms Catalog*, produced by the Russian Defense Ministry, is still required reading in Tehran, as it is in many other third world capitals, and the mail order business in high-tech weaponry flourishes.

In these circumstances, an effective global defense against missile attack must be regarded as a matter of the greatest urgency, not something to be researched at leisure in case the international situation becomes ugly. It is not only the dreadful consequences that would flow from the use of missiles armed with nuclear, chemical, or biological warheads, but the implications of their threatened use, that should disturb us. For the threat undermines the ability of the West to project power in defense of its allies and its interests. Would President George Bush have been able to construct the international coalition that removed Iraqi forces from Kuwait if those who joined up had immediately been at risk from Saddam Hussein's missiles? At the very least, the possession of ballistic missiles by an aggressor introduces new factors into the calculations faced by the political leadership when deciding how to respond.

European-American Synergy

The problem is at least as serious for those European states such as Britain and France that have the ability to project power as well as a long tradition of doing so. For obvious geographical reasons, the problem is likely to mature more quickly than for America. How can Europe's lack of interest in so obvious a threat be explained?

Is it because the emerging dangers of which I have spoken serve as an unwelcome reminder of just how dependent Europe still is upon America for its security—at a time when European federalists are trying to create a purely European defense? It would be the height of irresponsibility to allow such factors to take priority over the first and most important task of all governments—to protect the lives and property of their citizens.

How can we explain America's slow progress in exploiting its technological strengths through the deployment of a global system of ballistic missile defense? Is this in part attributable to outdated regard for the antiballistic missile (ABM) treaty, a relic of the cold war, which effectively makes U.S. vulnerability to a missile attack a legal obligation? Is the lack of enthusiasm shown by its allies a contributory factor? One of the signatories to the treaty no longer exists; and the world has changed in several other respects. But President Clinton reaffirmed his commitment to the treaty at the recent Helsinki summit.

The purpose of the treaty was to prevent the buildup of Soviet strategic offensive systems. It failed dismally in this respect, as Henry Kissinger, one of its principal architects, later acknowledged. But one can understand America's original reasons for wanting it. If, however, one of the principal features of the post–cold war period is missile proliferation—and no one actually denies this—why preserve a treaty that perpetuates the vulnerability of both parties to third-party attacks? Those in favor of keeping the treaty are apt to describe it as "the cornerstone of our stability." I think of it in the opposite light: as a major contributor to growing instabilities.

To conclude: a wider role for NATO and expansion to the east provides the best means for renewing—and deepening—Atlantic institutions and of giving new significance to a continued U.S. presence in Europe. Attempts to "politicize" NATO or to load it with responsibilities that would dilute its mutual defense capabilities or obscure its basic purpose must be resisted. Nor should we permit such factors to discourage us from asking inconvenient questions about the source of present dangers and the means by which these should be countered. This will require a collective response under American leadership to the urgent problem of missile proliferation, with NATO providing the practical means by which others may make a contribution. This in turn will necessitate some institutional adjustments, and in the European case, some revision of ambitions that, unless qualified, can only endanger the lives of European citizens. Above all, what is required is Western unity.

My optimism about the ultimate achievement of this goal rests upon faith in the values and resources of the civilization that we meet to defend.

PART TWO
Commentary

Commentary

The New Atlantic Initiative

CHRISTOPHER DEMUTH (United States): The New Atlantic Initiative was conceived and organized in 1995 to counter the complacency that had settled so quickly over the Western nations following the collapse of the Soviet Empire—a complacency, moreover, that seemed to consist in the abandonment of precisely those principles and practices that had just produced the epochal victory of the Western democracies in the cold war.

As the newly liberated nations of Central and Eastern Europe struggled to establish democratic legitimacy, market economies, and the rule of law, the governments of Western Europe seemed too self-absorbed to reintegrate them promptly and with alacrity into the Western community of free and open trade and the NATO alliance of mutual security. In former Yugoslavia, irredentist aggression and human slaughter were met by an irresolution and strategic confusion all too reminiscent of recent catastrophic security failures.

In Western Europe, noble impulses to forestall further national wars through closer economic and political ties had grown in the minds of some into a fantastic vision of a European superstate with tenuous connections to popular sentiment and democratic consent slouching toward Brussels. And in America a new generation of political leaders had emerged on the scene, many with dangerously thin memories of the provenance and purpose of the Atlantic Alliance or of the imperative of American leadership. On both sides of the ocean, preoccupation with domestic politics threatened to lead to international estrangement.

In that context, a group of European and American politicians, intellectuals, business executives, and journalists, convinced of the urgent importance of countering these trends and of a rededication to the defense, entrenchment, and extension of our Western inheritance of freedom, met in Prague in 1996. It was supremely fitting that the New Atlantic Initiative's First Annual Congress should be held in Prague—source and repository of some of the greatest glories of European art, music, and literature; victim of Nazism and communism and of Western irresolution; and scene of the boldest and most successful efforts of the past decade in rebuilding democracy, and free enterprise. It is equally fitting that our Second Congress should be in Phoenix—quintessence of the American West; source and repository of some of the most stirring creations of ancient, native, and modern American culture; scene of staggering natural beauty; and dramatic proof that free men and women can not only make the desert bloom but advance to the highest peaks of human commerce and civilization.

The Nature of Atlanticism

CHRISTOPHER DEMUTH (United States): Historians will long debate and hypothesize why it was that in the societies of Europe and America during the past 300 years there emerged a civilization that ennobled the human spirit and enriched our material circumstances to a degree utterly unknown anywhere else on the planet, anywhere else in all of recorded history. Our inheritance of the Greek tradition of human reason, shaped and elevated by the European Enlightenment, had much to do with it. The Christian insistence on the divinity or the dignity and integrity of every individual was crucial.

Competition between church and state for the allegiance of the common man and the consequent extension of the principles of freedom and open competition to the realms of science and commerce were also powerful elements in igniting the forces of progress. Whatever the causes, we can say with certainty that the practical consequences of freedom have been unprecedented. Nobel economist Robert Fogel goes so far as to say that the mastery we have achieved over our environment in the past 300 years of the West amounts to a new stage of human evolution.

In Europe and America, physical stature has increased a full 50 percent during these years, and average lifespan has more than doubled. Even people of modest means live today in circumstances that the kings of centuries past could scarcely have imagined. Such are the fruits of freedom. Today new technologies, new economic developments, and new spiritual awakenings in Europe and America intimate that the freedom-based progress of the past three centuries, far from having run its course, may be about to ascend to a new plateau.

JON L. KYL (United States): From Prague to Phoenix may seem like quite a distance, not only geographically but culturally. Take the cowboy. In the world of international diplomacy, to be a cowboy is to be a loose cannon, someone who is unpredictable and not a little vain and glorious. But I would like to point out other aspects of the cowboy mystique.

Cowboys, those horsemen of the Western plains, come down to us in American literature and legend not as an erratic sort of individual but as the steadiest of the steady workers—honest, unassuming, the soul of honor—and very good at what he does. While the cowboy is not a very educated man, he has an undeniable knack for dealing with different sorts of people. The cowboy could put a patronizing city slicker from back East in his place with three or four civil but sarcastic words. If an outlaw came to despoil the town, the cowboy would wield his six-shooter with the same ease and unanswerable effect. The cowboy and, in fact, all the adventurous men and women who came to the West in the past century had to pitch in and help create the conditions for civilization. They had to bring law and order to a harsh and beautiful frontier, so that schools, towns, churches, marketplaces, and eventually great modern cities could flourish. In Arizona and throughout the Southwest, we celebrate our forebears who made a new society take shape and who saw to it that freedom could coexist with order.

The American West and Western civilization were both built on deeply held convictions about the dignity of the individual and the protection of that individual by limited government undergirded by the rule of law. Our belief in individual rights and our resolve to oppose tyranny bind us, the members of the New Atlantic Initiative, together. Although the principal threats to international peace and stability may have changed over the past few years, the task remains the same: to promote the security, well-being, and expansion of the community of nations that respect the democratic rights of their peoples.

CHARLES POWELL (United Kingdom): In the case of the New Atlantic Initiative, I have always assumed that it is the initiative that is new, not the Atlanticism. Atlanticism is anything but new, and certainly does not need to be. Rather than talk too much about new Atlanticism, let us focus on how we can put the case for old Atlanticism in new and more telling ways and, thus, retain public support for it.

I distinguish between two aspects of Atlanticism. The first is operational Atlanticism: that is, the strategic partnership across the Atlantic between the United States and Europe, a partnership forged in war, both in World War II and in the cold war, but just as important in peace. It is primarily a military defense relationship, with NATO as its principal institution, although, of course, the relationship reaches beyond defense into politics and other areas. Then there is the second aspect, the more philosophical aspect of Atlanticism, which I see as an attitude that comprises a belief in the market economy, the rule of law and democracy, and a limited role for government. In short, it is everything represented by the expression *free world* or, as President Havel put it rather more elegantly, it is the essence of European and American civilization.

First, Atlanticism expressed through NATO has led to the defeat of communism in Europe and most of the rest of the world, and it continues to keep the United States engaged in Europe. That commitment, of course, does not end threats to democracy and to our security. Those threats have merely changed: the main ones do not come from the superpowers but from rabid dictators of relatively small countries, from distorted nationalism, and from terrorism. But I know of no better way to keep those threats at bay than Atlanticism.

Second, the broad ideals associated with Atlanticism are gaining ground in the wider world precisely because they have proved successful. It is in our interests to see that Atlanticism spreads because those principles offer the best hope of stability and prosperity for the largest number of countries and, thus, ultimately, for our own security. Likewise, the case for Atlanticism largely makes itself with values. Some might ask, What is the problem then? In fact, there are some challenges to Atlanticism, and we need to understand them to make an effective case for continuing Atlanticism effectively.

The first challenge is to demonstrate the relevance of Atlanticism to current events. It still seems to be linked in many people's minds with the past rather than the future. Atlanticism is associated with a cold war age, which now seems a long time ago. It recalls the British public's obsession with World War II films with titles like *Escape from Colditz* and *The Dam Busters* and other such dramatic titles. Atlanticism conjures up images of searchlights, guns, divided Berlin, and massed Soviet divisions. The idea has not made a complete transition to the post–cold war world.

Governments have not helped by moving slowly over the enlargement of NATO. By "going wobbly," to coin a phrase, for so long they have given credence to the argument that NATO and Atlanticism are outdated and rather hostile, inappropriate concepts and that it is rather shameful to be adhering to them still. It is important to the public image of Atlanticism to seem relevant to the modern world, unlike such defunct *isms* as communism. According to the simplistic view, *they* (the liberals) have given up communism, so *we* (the conservatives) can give up Atlanticism. This view also underlines another point that troubles me, the generational aspect of Atlanticism. During my years in the British diplomatic service, I saw the dominant mindset move from being naturally Atlanticist, naturally thinking of America first, then to being European, and then to being fixated on the European relationship—a trend with which I find myself out of sympathy.

It is alarming that the mission statement recently issued by Britain's new foreign secretary had no place among its priorities for Britain's relations with the United States. That relationship would traditionally have occupied first place in any statement of British foreign policy aims. Documents like this are not written by chance, and I fear it is symptomatic of the Eurocentric thinking that threatens to dominate Britain's new foreign policy establishment, as it had begun to dominate the old.

Very few things concentrate the mind as wonderfully as a good international crisis, which brings home to people in Europe the vital contribution that the Atlantic Alliance makes to their security and brings home to people in the United States the value of having reliable allies. I do not suggest that we should manufacture a crisis just to give Atlanticism the boost it badly needs, but sometimes a shock is salutary. Having successfully made the case for NATO enlargement, we now need to make the case for greater NATO activism beyond the present narrow definition of the NATO area. That is where the main threats to our security will arise, and that is where the Atlantic Alliance will need to demonstrate its continuing relevance.

Is there, perhaps, scope for action on the political level to widen support for Atlanticism? I was impressed in the 1997 Britain general election that while the Referendum Party achieved only a modest success, it did succeed in moving the European issue toward the top of the political agenda, forcing both the main political parties to commit themselves to a referendum on whether Britain should join a single currency. Single-issue groups can indeed have a telling effect on the political agenda, as Americans can surely attest. Perhaps there is more we can do in the New Atlantic Initiative to persuade candidates for election in our respective countries to adopt an Atlanticist commitment. Another effective way to demonstrate Atlanticism is to give it a stronger trade dimension, as Lady Thatcher has proposed in the past with a transatlantic free trade area. If we can convince people of the economic benefits of Atlanticism, we are well on the road to persuading them of its other benefits.

MAX M. KAMPELMAN (United States): Our purpose in the New Atlantic Initiative is to reinvigorate and expand the Atlantic community of democracies. Our premise is that it is in the national security interests of each of us that we nurture, protect, expand, and reinvigorate democracy where it exists.

As we prepare to exit the twentieth century, often described as the bloodiest in world history, we want to be assured that we marshal the will and the wisdom to realize our aspirations. These aspirations are attainable. With continued vigilance and determination, all should be enabled to enjoy the fruits of democratic dignity, free of fear and violence. The opportunity is there. Whether we have the wit, wisdom, and will to take advantage of the opportunity is yet to be determined.

Freedom House, where I have served as chairman emeritus, informs us in its authoritative annual report that more than 60 percent of the countries of the world, with about 55 percent of the population of the world, can now call themselves democratic or near democratic—the highest percentage in recorded history. That number can and should increase.

The communication age has opened up the world for all to see. The less fortunate are now aware that they can live in societies that respect their dignity as human beings. They know that such societies, which provide better health, improved sanitation, and adequate food and water, are only hours away from where they now live. They want that dignity and better living for themselves and for their children, and they do not wish to wait.

We cannot, however, be sanguine. While we have been steeped in evolutionary theory and hope that the human race is evolving toward a higher form of civilization, prudence urges us not to forget that the devil, too, evolves. It is in our interest that we take note of, recognize, identify, and resist the forces that stand in the way of our objectives. Some argue that our efforts to force democracy in other geographic areas is a misguided and doomed effort to transfer the religious values of our culture to other cultures not hospitable to those values. Our Western values, it is said, particularly by defenders of Middle East and Asian authoritarian systems, are unique to our Judeo-Christian culture alone. It is true that the modern idea of democracy originated in the West, but Judaism, Christianity, and Islam originated in the Middle East, and those ideas spread to all parts of the globe.

The ideas of freedom, however, need not be confined to Western Europe and North America. Westerners do not uniquely carry a democracy gene. We know that the ideology of the Enlightenment, which originated in the West, has established a bridgehead in all the non-Western civilizations. Young people of today's Japan, for example, are, in many ways, culturally closer to their American and European contemporaries than they are to their own grandparents. Our aspirations are universal. Let us mobilize that universal strength. A world community of democracies or a world congress of democracies should be our ultimate goal. As we enter the twenty-first century, such a challenge is worthy of our aspirations, our attention, our talents, and our energies.

MICHAEL ZANTOVSKY (Czech Republic): Atlanticism is a way of defining the essence of the political economy, culture, and spiritual features of our civilization, of things we consider so basic and so indispensable that they are worth fighting and even dying for. In the past, great civilizations have often perished not because of a defeat in a war but because they lost focus and clarity about a fundamental nature. In other words, they became deconstructed, first ambiguously, then clearly. The task of a new Atlanticism is to help prevent this from happening again.

BRIAN BEEDHAM (United Kingdom): What are we trying to create? In my view, we are trying to create a permanent replacement for three ad hoc single-purpose American-European alliances of the past century: the alliance of 1917, the alliance of the early 1940s, and the alliance that began in 1949 for the cold war. Each of these alliances came in response to a single crisis, and it succeeded. After a century of experience of discovering that Europe and America can work together on such a wide range of issues, we are trying to convert this experience into a permanent standing alliance.

We are doing what states did some hundreds of years ago when they discovered that it was not very sensible to create an army to fight a war and then disband it, and then create another army later on for the next war. They devised the concept of the standing army—we are now trying to create the concept of the standing alliance. This task is not easy because a group of states is obviously a different thing from a single state. It is not easy because there are differences of interest, some of them quite sizable, between Europe or some European countries and the United States and, indeed, among some European countries themselves.

It is not easy because in Europe, and particularly in France, there is a kind of cultural counterrevolution, a desire to believe that Europe is really a rather superior place and should assert itself and assert its separateness from the United States to recover the glories of its past history. For all these reasons, this project is not going to be easy, but I think it can be successful if we add to the continuation of the military alliance some structures and institutions in economics and some system of intergovernmental discussion and cooperation.

If we are successful in creating a permanent alliance of the democracies, what will it find itself doing in the next twenty to thirty years? In the first stage, the major crises are not likely to occur to the east of Europe. They are likely to happen to the south and the southeast of Europe, because that area contains the explosive combination of oil, many primitive authoritarian regimes of the sort that the democracies bypassed two or three centuries ago, and the widely misunderstood element of Islamic fundamentalism, along with the growing acquisition of weapons of mass destruction. I do not predict that an Islamic superpower will arise. But even without that, there are in this arc of countries to the south and southeast sufficient grounds for collision and explosion to make me believe that we need a standing alliance to withstand the challenge. In particular, Europe still needs America for that.

Somewhat later, China will almost certainly be a major power in the world. When China does emerge on a large scale, America will need Europe because maintaining a balance of power in East Asia will be very difficult for the United States. There is a collection of disparate allies in the area—Japan, Indonesia, and some Southeast Asian states—but I think Europe will be important as well. Therefore, this American need of Europe in that second stage balances the European need of America in the first stage.

JOHN O'SULLIVAN (United Kingdom): There are two aspects to Atlanticism. The first is the spirit of Atlanticism, the values of democracy and liberty, which are its essence. And there is, second, Atlantic policies, policies of governments on such matters as NATO and the transatlantic free trade area. And if we are to represent both, as we hope to do, then we have to recognize that many people who are united by the first, the values, will not always agree on every specific policy that Atlanticists are promoting at any one time.

Although, for example, NATO enlargement enjoys wide support, that support is not universal. And if universal support is absent on an issue so basic, differences will persist on other questions. We have to see ourselves as a broad church. We describe ourselves as bipartisan, but, in fact, that slightly understates it. We are in fact attempting to be representative of the debate in the whole culture. Just as in a single country people will differ violently on specific policies, nonetheless they are united in their concern for the welfare of the country. We expect a vigorous and lively debate, a debate that reflects the desires of all to ensure that the Atlantic world flourishes.

U.S.-European Ties

THOMAS J. DUESTERBERG (United States): I began promoting the idea of a free trade agreement between Europe and the NAFTA area several years ago. There seemed, at that time, to be a certain amount of political interest in the idea on both sides of the Atlantic. Prominent proponents of the Atlantic Alliance, including Lady Thatcher, Henry Kissinger, Brent Scowcroft, Newt Gingrich, at one point, Klaus Kinkel, and others had supported the idea. While it has lost a certain amount of momentum in the past few years, it is still a useful idea. Europe could realize major gains from a free trade agreement across the Atlantic. There, a continuing competitiveness problem is symbolized by a persistent high rate of unemployment and stagnant growth.

The approaches of the United States and of certain European powers in their movement toward a balanced budget form a contrast. In the United States, it is a political imperative. In Europe, it is a legal imperative for those countries that wish to join the Monetary Union in 1999. On the U.S. side, we have at least managed to get an agreement on paper to balance our budget, and the key factor in that was the surprisingly strong economic growth in the United States over the past few years. That growth has allowed us, miraculously, to find $225 billion in additional revenues, allowing the 1997 budget agreement between the White House and Congress. To meet EU convergence criteria in relation to public debt, the German government tried to revalue its gold stocks, and France transferred employee pensions to its list of public assets.

For Europe, a free trade agreement would force the process of restructuring that the United States has gone through during the past fifteen years. The discipline on corporate behavior imposed by free trade has led to a much more competitive environment in the United States than we had fifteen years ago. It might be easier in many European countries for that sort of restructuring to be accomplished if it was more or less forced on them by an international agreement.

It would be good for building the new transatlantic alliance to have an economic pillar. What better way to do it than to go through the process of a negotiation over a free trade agreement? While many have criticized this idea, though, as contributing to undermining the World Trade Organization, this argument, too, may be overdrawn. Nothing prevents us from negotiating a regional agreement that would be compatible with the WTO.

In recent years, the WTO has been dominated by the United States and the European Union. Nothing is accomplished without agreement between Europe and the United States. Two recent triumphs of the WTO, the information technology agreement and the telecom agreement, illustrate what can be achieved with the concerted action of the United States and the European Union. Those agreements were possible because the United States and Europe worked closely together, set out their priorities, and were able to bring other nations along.

Many tough issues loom in the future for the World Trade Organization. If we do not cooperate effectively across the Atlantic, some of those issues may spiral out of control. I would note the pending problem of integrating China and Russia into the WTO. At this point, I see little real cooperation between the United States and Europe on the terms of entry of those two major powers. If we fail to get the terms of entry right for those two powers, that failure could undermine the long-term legitimacy of the WTO. As matters stand now, China is able to play off the United States against the various European powers. What we have now appears to be a scramble for a minimalist agreement, a scramble for the bottom, which is all too typical of what happens in international organizations when China and, to a certain extent, Russia are involved.

In addition, difficult new areas like trade in services and intellectual property rights will be transacted over the Internet as its use becomes more widespread. We can make a transatlantic agreement open on a most-favored-nation basis, effectively allowing the globalization or multilateralization of any agreement that we reach. In that way, we could more effectively manage the WTO, a role that the United States and Europe have played effectively since the instigation of the Bretton Woods system more than forty years ago.

Finally, because the European and the American economies are well matched in trade and investment, it should be possible to reach a broad and comprehensive trade agreement. If we were able to do that, it would help counteract the tendencies toward more protectionism on both sides of the Atlantic. It is a doable agreement, and it would help challenge the centrifugal forces of protectionism from both Left and Right.

There is, however, no political consensus at the moment for such an agreement. In fact, the forces opposing further liberalization of trade in the United States, and I believe in Europe as well, have the upper hand at the moment. It will therefore take additional concerted political leadership to achieve an agreement, but it is well worth doing. We need that effective leadership, as shown by Lady Thatcher, Vice President Quayle, Henry Kissinger, Klaus Kinkel, Newt Gingrich, and others, and I would urge those leaders to continue the effort to articulate the reasons for such an agreement.

European Political Integration

CHARLES POWELL (United Kingdom): My experience of working in Europe in the 1970s and early 1980s left me convinced of three points.

First, Europe is and will remain obsessed with its internal construction at the expense of nurturing the transatlantic relationship. The internal will invariably take priority over the external. One sees that particularly in defense, where the focus is on creating ever-newer institutions in Europe, rather than designing Europe's contribution to solving security problems outside the European area. Second, the risk remains that we will create a Europe that is more competitor than partner of the United States not just in foreign policy and defense but also through, for example, the proposed single currency. Third, the traditional corporatism of continental Europe remains quite fundamentally at odds with the more free market liberal approach, which characterizes Atlanticism. Some signs of change are in the air, though. Daimler Benz in Germany, for example, has introduced the concept of shareholder value. But corporatism still remains a common approach.

The only serious attempt to set out an entirely different strategy for Europe, a strategy that would have given priority to free market economics and to the transatlantic relationship, was Lady Thatcher's Bruges speech of 1988, in which she set out the case for a Europe of cooperating sovereign states in contrast to the present plans to create a federal superstate. But sadly, in Europe, at least, the speech fell on deaf ears. Therefore, we have a rather eccentric situation in which the ideals of Atlanticism have increasingly wide appeal in Central and Eastern Europe and, indeed, further appeal in Asia. Yet on its home ground, above all in Western Europe, Atlanticism is in some danger of eroding. Perhaps we shall end up in the strange position of a more Atlanticist Asia than Europe.

RICHARD BURT (United States): What are the implications of an evolving European Union for Atlanticism? I think we are facing some very serious differences on both sides of the Atlantic over the role of the European Union, about how it will fit in with NATO and other Atlantic institutions, and what a European political union will mean for the alliance as a whole.

I see two very different models emerging. One might be termed the Franco-German model, which would entail the creation of a kind of United States of Europe. This model would require the existing governments of Europe to give up sovereignty, gradually but inevitably, to new structures and new institutions, many of which have not been tested and are arguably incompatible with some of the values of Atlanticism. The Anglo-American concept of a European Union, one focused on economic cooperation and a degree of political coordination, would stop far short of the kind of Europe-building vision that politicians in Bonn and Paris have embraced. It will be very difficult to get the Atlantic community to cooperate more effectively on issues like China in the face of a fundamental philosophical difference about how Europe should be constructed and how Europe should talk to America.

DAN QUAYLE (UNITED STATES): If in fact the European Union is a union that articulates the merits and the opportunities of free trade, that is one thing. If it turns into a fortress Europe, that is another. The policies of the EU are important and will affect the kind of relation the European nations seek with the United States, particularly on trade issues. Moreover, any diminution of the importance of the military dimension would be regrettable.

MARGARITA MATHIOPOULOS (Germany): Deputy Secretary of State Strobe Talbott has outlined a system of European Monetary Union that cements an open single market and sparks economic growth in Europe—and is also good for the American economy. Washington is therefore supporting European integration and the EMU process and has recognized that a strong Europe, including a strong European economy, is good for America in its efforts to share burdens. And Washington no longer makes distinctions between the Europeans. The once "special" relationship with Britain has in some ways perished. George Bush proclaimed in his famous speech in May 1989 that Bonn and Washington were partners in leadership. The Clinton administration works with the European Union as a whole, which is a new and very healthy process. No longer do we have Franco-German games versus Anglo-American games. We have all learned from the lessons from the past.

A Changing Balance of Power

MAX M. KAMPELMAN (United States): There are some disturbing signs in the world today. One is the lack of democratic unity in the face of state-sponsored terrorism. As highly sophisticated biological and chemical weapons increasingly appear on the scene, the danger of terrorism should not be underestimated by Europe or America. The

United States, which itself has a vacillating rather than an assertively consistent policy, urges sanctions, for example, to make Iran's leaders pay a significant economic price, while most of Europe prefers engagement, competes to help develop Iran further, and is eager, indeed, to meet Iraq's needs as well.

The disturbing weaknesses of our intelligence about terrorist activity can be demonstrated by the fact that no intelligence agency was apparently aware of the activities of the cult responsible for the 1995 sarin gas attack in Tokyo, even though its members—I learned to my dismay—owned a 48,000-acre range in Australia used to test biological agents on livestock, amassed an estimated $1 billion in assets, owned two Russian-made helicopters, had large stockpiles of chemicals, purchased a sarin production facility from Russia, and presumably had a small but worldwide membership. The terrorism danger is real—with biological and chemical threats but also with threats to our cyber infrastructure on which our economies are increasingly dependent.

JOSEF JOFFE (GERMANY): Being an adherent of the theory that alliances do not survive without a strategic threat, I wonder what can provide such threat. Brian Beedham has suggested China as one possibility and Islamic fundamentalism as another. Can that hypothesis be tested? China is hard to test because we have no experience with China. There is such an enormous strategic disconnect between Europe and China that I do not think that China can provide the basis for a Western European-Russian alliance.

The Islamic threat, in contrast, can be measured on the basis of twenty years of experience. The resurgence of Islam has taken a form that reflects explicit political ambition through terrorism, efforts to acquire weapons of mass destruction, and bellicose and aggressive behavior. All of that has been around for at least twenty years, and we, as NATO, have not countered those threats very well. For example, we have sold the technology of mass destruction to Iraq and Iran, with Germany as the great world exporter, although others have sold the technology as well.

Did we have a common front when Iran took over the American embassy? Did we have a common front when President Reagan decided to bomb Qaddafi's tents? No. I therefore conclude that while in the abstract Islamic fundamentalism should make a sufficient threat, the past twenty years have shown us that this threat is unlikely to galvanize Europe and the United States. In fact, the issue that created the most dissension of the past twenty years between the two sides of the Atlantic was precisely that issue

MARGARITA MATHIOPOULOS (Germany): Today, America is the only real superpower in the world, and today Washington is more powerful than it has ever been before economically, technologically, and militarily. As for the economy, it is evident that the Clinton administration managed in its first term to create 11 million new jobs, 2.6 million new jobs in 1996. The U.S. economy is booming in a very impressive way. As for technology, the United States is without any doubt first in the world. Asia tries to find a niche market, and we Europeans, especially we Germans, are far behind. As for the military, for the time being, the United States is the only world power. This shift in the balance of power manifests itself in the fact that America helps the Russians dismantle their nuclear missiles and carriers as well as in the U.S.-led revolution in the technology of armaments: think of the "cyber soldier," the theater missile defense systems, or the national missile defense system against middle-range and intercontinental missiles, new satellite systems, and so on. In addition, Washington plays the key role in every important foreign policy issue and trouble spot in the world and is more active and dominant than ever before.

Without President Clinton's role, the Hebron agreement would not be thinkable. Netanyahu and Arafat listen only to what Washington has to say and ignore completely the French voice. In Africa, the Americans are involved in Zaire, while in Asia, the United States keeps the balance between China, Japan, India, and the Little Tigers. In all of these areas and conflicts, America plays the most important role. That brief analysis makes clear, I believe, that the European-American relationship is not only good today but excellent. All those who said at the time President Clinton was elected that America would turn inward, become isolationist, and would care only about domestic issues proved shortsighted and wrong.

NATO Reform

CHRISTOPHER COX (United States): If we in Congress are able to consolidate our gains and victory in the cold war, to enlarge free Europe in a way that does not threaten Russia, and to protect ourselves against any long-term contingency that a Russian remilitarization or expansion might pose, then we are in a position to make much more of NATO and much more of our alliance.

We are now in a position to start cooperating on our relations with the People's Republic of China. At present, European and American cooperation—if it can be called that—on that subject reflects the old prisoner's dilemma: we are injuring ourselves to prevent our competitors from injuring us, but we do not have a concerted strategy for dealing with the rest of the world beyond our alliance. What is true of China is even truer of such states as Iraq, Iran, and Libya. I suggest that our response to the commercial and security challenges posed by those nations would be much enhanced by a well-functioning NATO.

WILLIAM E. ODOM (United States): For the past four years, I have been repeating Senator Lugar's phrase that NATO will go "out of area or out of action or out of business." And Bosnia is the first test case of out of area. Presumably, this effort should help breathe a new sense of purpose into NATO, but for that to happen, we must have some honesty in advertising how long U.S. forces will have to stay in Bosnia. I see a serious risk of crisis if Secretary of Defense Cohen really means what he has said about an early withdrawal and if those in the Congress hold him to it.

The president, in my view, made a serious mistake when he announced his decision to send troops to Bosnia. He made very persuasive arguments about why our interests were strategic and what it was we needed to do but then set a one-year time limit. Why would those arguments not be relevant in a year if the peace had not been fully established? Therefore, I think it is high time that both he and the Republican Congress found a bipartisan formula for making the commitment fairly long term. If they do not, I think the alliance is headed for a crisis.

RAINER SCHUWIRTH (Germany): If we want to maintain or achieve consensus in the future in this alliance, all of us will have to accept that NATO cannot be the playground for individual national interests. We should not try to make NATO the machinery for everything only because other institutions or organizations fail from time to time but because we should use NATO for what it is. NATO actions should reflect the highest common denominator possible. Because, however, this will not always be the case, this is one argument for developing a European security and defense identity within the alliance. I stress the words *within the alliance* to emphasize that the objective should not be to split the alliance but to achieve complementary purposes.

The core functions of NATO as defined in the NATO strategic concept of 1991 go beyond collective defense. They include, for instance, NATO's function as the forum for security and defense-related dialogue and consultations. That role entails recognizing the need to consult before individual initiatives are taken, which is easily said but in reality sometimes very difficult. In this respect, I would also like to see NATO do more to prevent crises than simply reacting after they have emerged.

What should we do in the near future? We should clarify the purpose of NATO enlargement, for example. As I see it, NATO should have a broader mission in maintaining security and stability in Europe and, of course, promoting American interests in the process. NATO has always been a military and a political alliance to promote shared values, and it should remain so. The old mission of keeping the Americans in, the Germans down, and the Russians out is rather obsolete. NATO's threefold mission now is to enhance Western cohesion, ensure Central European consolidation, and entice Russian cooperation. All three goals must be pursued vigorously, even if some of us are skeptical about Russia's becoming a very constructive partner.

At the risk of giving ammunition to critics of enlargement, I note that some of these old-timers used to echo the Communist Party line when speaking of NATO as "the main danger for world peace." While officials in Washington and elsewhere have changed their views, and some even out of conviction, the conversion of Central European military, diplomatic, and intelligence personnel assigned to NATO should be carefully scrutinized. These next several years should be also used to help the new democracies establish a more reliable system than they now have of protecting NATO's secrets.

RICHARD N. PERLE (United States): We must maintain the effective military strength of the alliance, and that increasingly, in my view, will require investment in a variety of advanced technologies with the potential to transform radically the way we fight wars. That technology is now largely the product of the civilian sector. It is widely available to those who choose to adapt it to military purposes. If we do not find the means to modernize our military forces in this way, no matter how well we deal with the political challenges, we will not have the underlying military capability that has always been the foundation of this alliance.

Means and Rationale for NATO Enlargement

MARK P. LAGON (United States): Does the United States have a clear strategy regarding NATO expansion? In the view of Secretary of State Madeleine Albright, the main purpose of NATO is to facilitate the consolidation of democracy in Central Europe. That view is regarded as consistent with a legacy of consolidating democracy in West Germany, Portugal, Italy, and elsewhere in the past.

But, obviously, one of the main reasons that Central European nations want to join is their concern about a potential threat; and if there is any consistency with the past, NATO will remain a collective defense organization, that is, an organization concerned with threats outside its membership. And frankly, that threat, while not likely, is most likely to come from Russia. Therefore, I do not see a clear notion about the main purpose of NATO. Both collective defense and the consolidation of democracy are important, but which is the main one is unclear. Some of that lack of clarity has profound repercussions.

The American administration is ambivalent about basic questions: do we include Romania and Slovenia in the first round of NATO expansion, along with Poland, Hungary, and the Czech Republic? Ironically, the strength of the candidacy of Romania and Slovenia and their very fitness for becoming members are perhaps regarded as reasons for delay because they can then be touted as really good candidates for a second round. In that way, we avoid such troubling questions as whether we let the Baltic nations in. The administration is confused about what the precise priorities are.

JEREMY D. ROSNER (United States): In the first instance, the U.S. strategy is to have an enlarged alliance. In the second instance, it is to have an enlarged alliance that enlarges not just once but more than once. NATO enlargement is thus not a one-time affair but an open door, as President Clinton has said. That is the core of the strategy. Those who believe that the strategy is not clear enough and who yearn to hear more exact details are not likely to do so until we get much closer to the actual admission of new members, and that uncertainty will be a frustration. As to the threats that the new NATO will face, I was struck by Mark Lagon's comment that the threat is all Russian, in contrast to Paul Wolfowitz's comment that the threat is *not* Russian.

We cannot assume a unity of opinion on the definition of the threat in either the executive or the legislative branch or in either the Democratic or the Republican Party. I do not think there is agreement. Nor is there a consensus on what will create a balanced process that will fulfill the pledge that we have made of a continuing process of enlargement. I think it is a mistake to suppose that uncertainty about which countries are to be admitted demonstrates a lack of strategic clarity. The strategic clarity has existed for four years. People have doubted it, but it is about to be demonstrated. When the decision on which countries are to be admitted becomes known, the underlying geostrategy will be clear.

MIRA BARATTA (United States): Despite all that has been said, there is a lot of confusion. Those from Central and Eastern Europe who do not seem to be part of the first wave of entry into NATO are uncertain what the criteria are, how the decisions will be made, whether the process will just be political horse trading, or whether there really is a geostrategy rationale.

OJARS KALNINS (Latvia): Many who support the NATO enlargement process envision the creation of a united, undivided Europe. It follows that the process could only stop once that goal was achieved, once Europe was successfully united. The only question then is, Who is part of Europe? Does that include Russia and beyond? That question has to be answered.

Those of us who are diplomats have to deal with some practical issues, and NATO enlargement is something tangible. I want to raise some questions of concern to us. First, will there, indeed, be a second wave of NATO enlargement? From what I have heard in Brussels and Washington, there seems to be a question about that. The debate in the United States appears to be over whether enlargement should take place at all. And I assume the answer to that question will be positive, but that applies only to the first wave. Will there be support for the second wave?

Second, if there is another wave, when does it occur? It matters significantly whether it occurs ten or fifteen years from now or whether it occurs two years from now. And what is of very direct importance to my country and many others is what happens in the meantime. What will be the relationship between us and NATO? Regardless of whether we are admitted to NATO in two years, five years, or ten years, we have to deal with immediate security concerns.

DOUGLAS J. FEITH (United States): Strategy, I suggest, is not a question of our selecting who should be in NATO now or later. Strategy ties into the question of why. How would we explain a strategy to a typical American legislator, for example—a senator who does not pay a great deal of attention to foreign affairs? How do we explain the point of NATO expansion and why the United States should undertake additional obligations, duties, and investments?

As Paul Wolfowitz said, NATO has several purposes. One purpose is to function at the military level and perform as a military alliance. We therefore need to know the nature of the current military threat. Beyond that, though, it has a broader function of promoting democracy, shoring up democracy in Central Europe, and influencing, to the extent of its power, events inside Russia, whatever positive trends there that may bring Russia into a more democratic government and greater cooperation with NATO. Those issues get to the heart of the rationale for NATO expansion. They force us to address the question of why there is a NATO at all now and what its purpose is. If we had a stronger sense of purpose for the organization, then the criteria for admission of future members would fall into place naturally.

MR. ROSNER: NATO's core purpose is unambiguous: the collective defense of NATO soil. I do not know where the notion that somehow the promotion of democracy has eclipsed that purpose comes from. It does not come from statements of the Clinton administration.

The preeminent purpose of NATO has always been to defend its own soil. The issue, then, is what the threats are to the soil right now and what other purposes NATO can serve—and how to balance them. It is a mistake to think that one geostrategy oriented toward one place will define everything from force structure to which new members to admit.

The Clinton administration sees NATO enlargement as an opportunity to extend the process of democratic evolution to include virtually all Europe, creating a buffer against a range of challenges from the east and from the south and consolidating democracy internally. Over a long period of time, we envisage a dynamic process that will bring all Europe's new democracies into a cohesive whole, whether in formal NATO membership or in some other form of partnership.

BRUCE P. JACKSON (United States): Clearly, the criterion for new admissions is the same as it has always been, and we are admitting new members for the same purposes we always have. We reunified Germany and brought Spain into NATO for the same reason we are bringing these other countries in. In my view, we should claim victory for the alliance, and we should be talking about the nature and significance of that victory.

The first characteristic of that victory is that it is progressive, so that whether Romania gets in or Hungary gets in is incidental to the claim of victory. And to judge whether some people are more deserving than others is the judgment of Paris, and that approach should be rejected. The second characteristic of this victory is its conclusiveness. We are asserting—and this is something that we should not be ashamed of asserting—that this is a victory of values. We are talking about freedom, and arguments over technicalities and costs are distractions. The final characteristic of this victory is that it is not contingent. It does not depend on who is in the White House or the political complexion of Congress. We have been working toward this victory for fifty or sixty years; it just happens to be occurring at this juncture. And if this Congress can do anything, it can affirm the central principles underlying that victory.

VALOIS V. PAVLOVSKIS (Latvia): Those of us in the Baltics wonder what will happen to us. Throughout modern history, we have been victims of betrayal. Europe has always betrayed us—we have always been a convenient pawn in the play of power politics. Some want to claim victory for the NATO alliance—I say a victory for what? For a united Europe? For progress in Europe? But what about the Baltics? Will there be a second round of NATO enlargement? When will that be? And what are the criteria? Perhaps those from Great Britain, the United States, Germany, and the other big countries would not find those questions important. But to us who have been fighting for survival for the past fifty years those are very important questions. The Baltic

States face a question of survival. We deserve an answer from Europe.

CHARLES GATI (United States): In my view, democracies, unless they face a major opponent, do not have foreign policy strategy—certainly not a single one. Perhaps the best way to answer the questions of those in search of that single one is to understand the diverse coalitions of those supporting NATO enlargement and to explain their very different reasons for doing so.

The pro-enlargement coalition does not endorse a single strategy. The issue is not Central Europe versus the Baltics. When enlargement will reach Latvia I do not know, but neither the Clinton administration nor the task force I participated in led by Senator Lugar intends this to be the last enlargement of NATO. When the next one will happen, I have no idea. Democracies do not plan that way. Probably other states do not either, but certainly democracies take things as they come, step by step, and so no one can truly predict when the next enlargement will occur.

MACIEJ KOZLOWSKI (Poland): My concern is that there is no strategy about what should be done between the decision to admit new members into NATO and the actual admission of the first wave. The interval, we hope, will be two years, but that may be prolonged. Life is unpredictable, and parliaments are even more unpredictable than life. With sixteen parliaments, some unexpected event—an election in one country, some internal debate on some very intricate issue in another—might postpone the whole issue. A two-year interval might work out, but if it is a longer time, then we in this first group will be in a kind of limbo that could create a problem. And problems with the first wave might create problems with the second wave. Still, we are on the move; we have begun the process.

We also need to maintain NATO cohesion. It is possible that its cohesion will become diluted, so that the new NATO becomes a forum for discussing problems, a kind of bureaucracy instead of a fighting machine. NATO was created to fight a war or to prevent a war. It has to be so strong that no one dares to attack it. And it has to be kept that way. It is easy to maintain an alliance when a clear enemy confronts you—with the enemy comes the cohesion. Without a clear enemy, it is very hard to keep all NATO countries working together in the very special sphere of defense planning and defense preparedness.

We would like the second wave, and even the third wave, to happen as soon as possible because we would like to include as many other countries as possible. We have the same strategy as Germany had: that is, we do not want to be the most eastern country within the alliance. Nor do we want a gray zone in Europe—Poland is now a gray zone. The countries not in the first wave will become part of a gray zone. The shorter the period a gray zone exists, and the smaller its size, the better for all our security, both that of the old NATO members and that of the new ones.

I foresee not so much a positioning of who is on the short list and who has the best claim but questions that are quite different and have not been fully discussed yet. There will be a close examination of NATO's willingness to act. What does it do? What tasks is it willing to take on? What happens in Bosnia? What happens on other fronts? Some have raised the issue of whether NATO is willing to act coherently against China. Those are the questions that people will look at in deciding what NATO's future will be about.

Questions about who is willing to pay will also arise. All these issues, much more than some set of entrance criteria, will determine what the next rounds look like. What happens will be based on the experience of the alliance and on what happens in real life situations. I am convinced there will be other rounds of NATO enlargement. It is useful to look at 1949, the year in which the Washington Treaty was signed by the original NATO members. Once that decision was made, the subsequent decisions in 1952, 1955, and 1982 to expand NATO were uncontroversial. I also think that it is useful to avoid regarding NATO as if it were a stockade: when the doors close, everyone who is left on the outside is vulnerable. It has always been the case that NATO has cast a shadow, a penumbra of security broader than its own borders. It is foolish to think that states not in the first round will somehow gain no benefit from NATO expansion.

JOSEF JOFFE (Germany): Recently, a watershed was crossed after several years of bargaining: Russia finally bestowed its approval on NATO enlargement, and I found it like watching an absurdist drama, something by Ionesco or Beckett. Why? First, this was the first time in the history of alliance politics when one group of nations asked another nation whether it might be permitted to add more. The textbooks contain nothing like it.

Second, the alliance took in those nations that do not need it and left out those that do, at least from a strictly strategic point of view. Of the three that are in, two do not share a border with Russia, nor would I think that they feel particularly threatened. In contrast, the four nations that have the most serious security problems in Europe—the Baltics and Ukraine—are out; nor do I predict that we will ever let them in.

Third, we have accepted the one nation that has been and in a way remains the raison d'être of this alliance—namely, Russia, the heir of the Soviet Union. But, of course, the absurdity in the NATO play is easier to crack than the absurdity in *Waiting for Godot* because with the demise of the great strategic threat that gave rise and sustenance to the alliance, NATO has lost its reason for being, like every other alliance in history that won a war. And what is more, it could not quite bring itself to say the obvious thing: that there is still life in the old lady because there was, and is, still a lot of life in the old enemy.

BRIAN BEEDHAM (United Kingdom): Who can and cannot be a member of the alliance? I think any country with a solidly established record of democratic practices built up over ten or fifteen years, sharing the same broad collection of interests as the existing members of the alliance, should be allowed to become a member, and I would certainly include the Baltic States in that. I would prefer not to experiment with different classes of membership, which I see as rather dangerous. It is the easy way out for nervous expanders. Membership in NATO should clearly mean what it says and carry the guarantees that go with it. Then I think we should invite in any country that passes those tests.

WILLIAM E. ODOM (United States):Most of us who have been arguing strongly for NATO expansion have not been fully candid on the inevitable problems that will come from accepting new members. We focused primarily on allaying the fears of those who worry about the problems. We know that disappointed applicants, for example, will face both domestic and foreign policy problems as a result of not being allowed to enter as part of the first tranche. We should also recognize that some new members entering the alliance will probably embarrass us on certain occasions. A recent spy case in Poland is an example. Other such cases could arise.

One that concerns me is that the old officer corps and the large bureaucracies are far too big and still too Communist in their indoctrination and training. From a professional military viewpoint, it would be much wiser to demobilize those organizations totally and start over almost from scratch to build anew. The aim must be to have effective, smaller, more modern military forces. But the new member governments may find it politically difficult to retire those people and to cut those bureaucracies back; instead, they will squander money that otherwise could be used more effectively.

We can expect that the agreement between Russia and NATO will be used to justify Moscow's hegemony over the Baltic States and Transcaucasia and will present other difficulties. Those are some of the problems that will arise. Shouldn't the proponents of the expansion take preemptive measures by openly anticipating those problems, pointing out that they will arise and that they are merely the price of the expansion, problems to be faced and solved, not reasons for refusing to expand? If we do not do this, won't the opponents of expansion seize every such problem as evidence that they were right and create political difficulties for us in the future? That situation would certainly not contribute to the deepening of the Atlantic Community. The proponents will be left with only counterfactual arguments about even greater problems that would have arisen if NATO had not expanded. We are making a tactical mistake if we do not take the initiative and try to preempt some of those arguments.

My third concern has to do with the future of NATO's military capabilities. I was pleased that Lady Thatcher raised the topic because I think it tends to be neglected. In making a case for expansion, I have frequently pointed out that NATO faces no imminent military threat from Russia. Rather it faces trouble-making, mischievous diplomacy from Moscow.

Disturbingly, some people in the U.S. government are speaking of making NATO more political and less military. Is that really a proper aim? In reality, NATO has been important politically precisely because it was militarily strong, and I see no reason for that to change in the future. Today, U.S. ground forces in Europe are vastly overstretched. A solid military case could be made for putting one or two more U.S. divisions in Europe, although I know that this would be extremely difficult politically, if not impossible. At the same time, key European states have been cutting their militaries fairly drastically, while some members of the U.S. Congress talk about European states sharing more of the military burden. I see problems here in the military area that, if not addressed, may spell difficulty for the task of deepening the Atlantic Alliance.

These three sets of concerns are hardly exhaustive. As I contemplate them, however, I am disturbed that we may be entering a period of enormous stress for the alliance because of a false sense of confidence and a misleading sense of euphoria created by deciding to expand the alliance and naming a few new members but not recognizing all the other problems that we face. If there is to be lasting substance to the efforts of the New Atlantic Initiative, we have a huge amount of work ahead of us.

DAN QUAYLE (United States): NATO enlargement is long overdue. It should have come sooner rather than later. But let us be realistic as we look at the NATO expansion and ask ourselves some questions about it. First, will the three countries entering NATO have full rights and responsibilities, or will we allow the Russians to set conditions on their membership? They must have full-class citizenship when it comes to NATO. I do not know the details of any such side agreements with the Russians, but we will find that out as the treaty is submitted to the Senate. It takes a two-thirds vote for ratification, and there will certainly be discussion on the terms of the treaty—it is not a *fait accompli.* I presume in the end it will pass but not before vigorous debate.

Questions about the costs must be resolved, because we need to know how much more money we will have to spend. Most important, is it in our national interests to do this? My answer is yes, of course. But will expansion stop with these three countries? Probably not. What about the Baltic nations? They are certainly the most vulnerable right now, and shouldn't the Baltic nations at some time be considered for membership? I dare not even bring up Ukraine now, but at some time down the road we might consider that country if we are successful with this first wave of NATO expansion.

We also need to keep in mind that the core functions of NATO as defined in the NATO strategic concept of 1991 go beyond collective defense. They include, for instance, NATO's function as the forum for security and defense-related dialogue and consultations. That role entails recognizing the need to consult before individual initiatives are taken, which is easily said but in reality sometimes very difficult to do. In this respect, I would also encourage us to do more to prevent crises rather than simply reacting to crises that have already emerged.

Costs of NATO Enlargement

JEREMY D. ROSNER (United States): The U.S. administration has estimated that the total cost of NATO enlargement—costs for current allies, new allies, and investments in interoperability—at $27–35 billion over about thirteen years. There are higher and lower estimates, of course. But whatever the number, these are serious figures, and they will generate serious debates.

When the Senate appointed a group of twenty-eight members to work with the administration on the NATO enlargement ratification process, I had expected questions about rationale, about whether enlargement would dilute NATO's military purposes, and especially about the NATO-Russia Founding Act. But, in fact, those issues have provoked very little interest. Rather, discussion has focused almost entirely on the cost of enlargement. For me, the defining moment came when a conservative Republican senator asked several perfunctory questions of Strobe Talbott on Russia—and this on the day after the Founding Act was released—and then said, "We seem to have settled the Russia issue. Let's talk about what this costs." And that was the end of the discussion about Russia. This cost issue will play an enormous role in the political debate over NATO enlargement both here and abroad. As we prepare for that debate, I think it is worth noting that there are at least six different cost issues.

First is the question of what a larger NATO needs to buy to fulfill its security missions. That, in turn, raises questions about the nature of the threat and the proper military strategy, both recurring themes in the cost discussion. Second are the costs arising from the need of the new members to upgrade their own national militaries and achieve interoperability. That demand comes at a time when those states are grappling with hard issues about economic and social policy reforms. Third are the costs for America's current European allies arising from their need to improve their national militaries. Since the beginning of this decade, NATO's entire military strategy has been premised on power projection, and yet at this point, the United States is far ahead in its ability to achieve mobility and power projection. Many of our allies will need to make investments to reach the standards that NATO has already set for itself in past years. This requirement, obviously, comes at a time when states in Europe are scrambling to meet convergence criteria for European Monetary Union.

Fourth are costs of interoperability. NATO will need to decide what it spends in its common funded budget for improvements on infrastructure, and this will raise questions not just about military strategy but also about difficult issues of alliance politics between, for example, southern and northern members. Fifth, it will be argued that enlargement diverts resources from spending on other military needs—procurement, security efforts in Asia, efforts to raise the quality of life for the troops, and the like—at a time when defense budgets are, at best, stagnant but more probably declining. Sixth is the matter of burden sharing, the key to the debate in the United States. The issue is not so much a concern over the absolute dollar figure but a concern that Uncle Sam will be taken for Uncle Sucker, as the senators often put it, with the United States paying a disproportionate share of costs, as current and new allies try to free ride. The issue arouses such strong feeling that senators have begun talking about legislating cost-sharing percentages.

ROBERT MROZIEWICZ (Poland): Membership payments differ from country to country, depending on the country's size, its gross national income, annual defense budget, and other measures of wealth. Poland's payment may be expected to be somewhere between that of Spain and Turkey. According to available data, in 1995, the sum total of the three NATO budgets amounted to $1.766 billion. Assuming Poland's contribution at some 2 percent, the actual payment would amount to $35–40 million annually.

In accordance with suggestions found in the study on NATO enlargement, the first priority assigned to the Polish Ministry of Defense was achieving compatibility and interoperability with NATO forces. Toward this goal, the following tasks have to be implemented first: (1) integrating the Polish command system with NATO; (2) achieving full compatibility between Polish and NATO telecommunications systems; and (3) ensuring compatibility of the Polish air defense system with that of NATO countries.

Poland will therefore have to create an integrated air traffic control system compatible with that of NATO, integrate its military and civilian data processing systems, and connect with the NATO air defense system. It will need to adopt new flight safety measures, gradually introduce modern weapons and NATO standard equipment, upgrade the country's military infrastructure, and modernize its air force. The gradual introduction of modern weapons and NATO standard equipment and upgrading the country's military infrastructure are, by far, the most costly parts of the process. Only upgrading command communications and air defense and air force modernization are integral to acceptance into NATO. Tasks considered indispensable to Poland's national defense, regardless of whether Poland is a member of NATO, should not be counted among the costs of joining the alliance; the expenditures listed above would probably suffice to enable the Polish armed forces to join the alliance. Accordingly, the preliminary estimates made by the Polish minister of national defense call for expenditures amounting to approximately $1.26 billion.

The entire program of restructuring the Polish armed forces will last until the year 2010, which translates into an average annual spending of some $0.4 billion or 3.3 percent of the Polish Defense Ministry budget in 1995. Let us remember, though, that the major part of the cost of achieving interoperability will have to be paid in the early stages of integration, by the year 2000. As a result, the annual funding requirement will be uneven throughout the period under discussion. Therefore, the realistically estimated costs of Poland's integration into NATO do not exceed the country's budgetary capabilities, although one can venture to predict that the ultimate costs to current NATO members will prove considerably lower than those projected in some U.S. studies.

RÉKA A. SZEMERKÉNYI (Hungary): The handling of the cost issue is an indication and a measurement of the credibility of the political will of NATO's current members and of those applying for membership. The significance of the cost factor goes well beyond the first tier of invitation for NATO membership and of accepting the first new members. Resolving the issues related to costs will be one of the determining factors in the success of the entire concept of enlargement and will also be a determining factor in the timing of the subsequent tiers of enlargement and the pace and scope of the next rounds, if there are any.

Although we can say that this is a critical issue at a critical moment and that we are moving from theory to reality, we have to acknowledge that it is a fragile reality. Talking about the cost of NATO's enlargement is almost like walking on eggs. In practical terms, it represents an adventure into the unknown.

The conclusion of the Hungarian Ministry of Defense experts who worked on this analysis, as well as those who worked on it at the Hungarian Atlantic Council, is that these are manageable costs economically and also that they are salable to the Hungarian public. Will this be the end of the story? Could we say now that, since we have figures that seem to be approved by the United States and that are financially manageable in Hungary as well, we have a done deal?

There are other factors at work in shaping attitudes toward increasing the defense budget. On the positive side is the domestic political consensus in Parliament among all the parliamentary parties. While there is concern about the costs of national defense, we also recognize that security and stability are the bases for economic development and political democratization. Although the Russian economy does not allow much adventurism in the short term, this break will not last forever. Even now, the Russian black market has started to creep into Central Europe and exercises a significant threat.

What factors constrain support for increasing Hungary's defense budget that would allow for NATO's enlargement? Significant difficulties arise from an economy in transition and impose a strict fiscal limitation. We could say that we can see light at the end of the tunnel, but it is not clear that everybody really does see the light there. Whatever the cause, little was done between 1994 and 1996 to raise public support, and so two years were lost. The government's intention to hold a referendum on the issue is also uncertain. While the government pledged in 1994 to hold a referendum on NATO's enlargement, it currently seems to be trying to back out of this promise. Its intention, though, is not clear.

Moreover, there is an unclear commitment to raise the defense budget, which is currently at a devastatingly low level. All this leads us to conclude that the parliamentary role in supporting an increase to the defense budget and Hungary's admission to NATO is crucial to the process. The parliamentary opposition parties have to be on guard continuously. They have to make sure that the commitment of the government will be clear and understandable and that the Atlantic idea and Atlanticism will take root. Domestic leadership is therefore crucial.

K. ERIK TYGESEN (Denmark): All governments must be absolutely sure that they read their parliaments correctly on the issue of NATO enlargement. The worst thing that could happen to us would be to negotiate an agreement that we could not get through sixteen national parliaments. I do not think ratification will be difficult in Denmark because in Denmark it is very easy to explain to the population how enlargement will enhance our security. But European countries farther away from the center of the action, as well as America, may find it much more difficult.

During this crucial period, it would be well to keep a constant eye on what is going on in Europe, where in addition to NATO enlargement countries will be negotiating the enlargement of the European Union. Those negotiations will cover such difficult terrain as agricultural policy reform, structural reform, and financial policy reform. A number of countries, especially in the southern part of the EU and the alliance, will have a very difficult situation with their parliaments because they will stand to lose on all three counts. They will get less from their agricultural policy, they will get less from the structural funds, and they will have to pay more through the financial system—and all of that for EU enlargement. These issues are being debated at the same time we are ratifying the NATO enlargement, which will also cost some money.

KALEV STOICESCU (Estonia): Our belief is that potential members of NATO should not be forgotten in the concerns over the first wave of new admissions. Otherwise, the difference in the quality of defenses of, for instance, Poland and the Baltic States would increase even more. Differences exist now but would grow greater after the entrance of the new members.

I do not raise this issue because my country is suffering from economic difficulties—on the contrary, we are witnessing an economic boom—but for other reasons. Because we were not party to the Conventional Forces in Europe talks, we did not inherit any armaments. We do not possess a real military industry in our country. If no political support is extended to my country and the others that fail to be admitted as part of the first tranche, this will be interpreted as a clear sign that NATO is not serious about continuous enlargement.

MACIEJ KOZLOWSKI (Poland): I am a little uneasy with the way in which the debate on costs is taking shape. As I see it, we are now being punished for the dissolution of the Warsaw Pact because the United States had carried the burden of defending Europe for forty-five years. No one was asking then how much money had to be spent, although it was obvious that the money had to be spent. With the collapse of the Warsaw Pact, the United States defense budget diminished considerably. In our recent study, we included a calculation about the size of the so-called peace dividend. The figure amounted to billions of dollars. But now, when there is no Warsaw Pact, no imminent threat, there is a discussion about cost sharing. The danger is that we are overlooking the crucial point, which is that it will cost much, much less than it cost before.

To put matters in perspective, I think a study should be done to determine how much it would cost *not* to enlarge NATO, that is to say, a study on how much the U.S. taxpayer will have to spend if our armed forces remain in their cold war mode. The calculation also needs to reflect the fact in the post–cold war era that NATO will be stronger with the inclusion of the new members than it would be without them. What is needed is a shift in the public discussion from the figures provided by experts to the political question: what kind of security is needed in the new environment of Europe? It will then be possible to start to calculate realistically the costs involved rather than on the basis of assumptions made in the context of a totally different situation.

We have done our homework because we understand that we have to be a reliable partner: NATO has to be coherent, and it has to have credibility. Therefore, our armed forces have to be compatible because otherwise we lose credibility. But credibility is all that is needed. After all, the forces are not going to fight tomorrow. We must make our armed forces credible so that no one dares to attack them, and that is all that we have to do.

MR. ROSNER: Are the calculations about enlargement purely arbitrary, as some participants have suggested? In a way, they are. In another sense, they are not. If the process of enlarging NATO is viewed as purely political, then obviously they are arbitrary. But nobody sees this process as purely political. NATO is a security and defense alliance; it is not just political. It has to have a clear-cut military purpose to it. The number-one priority is interoperability, and that requires absolutely clear definitions.

On the question of the second wave and when it will be, no one knows. But this I do know: both the speed and the enthusiasm with which it is greeted, at least here in the United States, will depend, in part, on the experience of the first round, including the experience of practical realities learned on the ground and the record on burden sharing. If politicians and the public conclude, after the first group has come in, that there has been genuine burden sharing, I think they will regard that as a model for the future. But if they believe that they were saddled with a disproportionate share of the cost, prospects for the next round will receive less support.

The NATO-Russia Founding Act

CHRISTOPHER COX (United States): The good news is that the Founding Act is not binding on members of the alliance, including America or the U.S. Congress. Moreover, to the extent that it embodies bad ideas, we can fix it. Even better, Congress is set on this objective. We are working to turn these lemons into lemonade.

MICHAEL ZANTOVSKY (Czech Republic): Although I am confident that NATO negotiators acted in good faith and with the interests of NATO and the interests of our countries in mind, the overwhelming desire to have an agreement may have produced something that will not be as easy to live with as we would like. The document is marked by ambiguity to such a degree that I am tempted to coin a phrase, *deconstructive ambiguity*. The ambiguity starts with the name: NATO wanted a *charter;* Russia wanted a *treaty*. What we are getting is a *Founding Act*.

It is not immediately clear what it is exactly that the act is founding. Ostensibly, it is founding a cooperative relationship, something that most of us sincerely desire. But even in "diplomatese," to speak about founding a relationship is somewhat unusual. More often, we speak about founding something slightly more palpable such as a structure or an institution. And I believe that the name in this instance is a clear reference to the final act of the Helsinki Conference of 1975—a process that gradually developed into an institution, albeit not a very robust or effective one, now called the Organization for Security and Cooperation in Europe (OSCE). It is no secret that the Russians have for a long time envisaged a European security architecture based on the OSCE and that they, at times, advocated the idea of transforming NATO into a larger structure based on that institution.

The ambiguity does not stop here. While NATO has provided political assurances that seem reasonable in the absence, but only in the absence, of an apparent security threat, the Russians seem to claim that they have won a binding agreement. Some leading Russian politicians have already said that the act will go through a ratification process in the Duma, thus making the act a part of international law. A diplomat would obviously ask, Who will be the other parties to the ratification? Where will the ratification instruments be deposited? But such questions do not seem to mar the happiness of many Russian leaders about such a stroke of luck.

The idea of an agreement that one side sees as binding and the other side sees as merely a statement of intentions would give even the most cynical diplomat the shudders. The consultative body that the agreement provides for is another source of ambiguity. The body will discuss any NATO policy issues at the whim of any of its members. Although this arrangement should not give the Russians the right to veto or codecide any of the issues, it certainly does look like a foot in the door of bearish proportions. It would be logical, if only for the sake of symmetry, to ask for a seat for a NATO representative on the Russian National Security Council. Such a seat was not offered, nor was it apparently asked for. The time element is another source of ambiguity. The council will be established immediately, while the moment for the new members to join is still two years away. Among the issues the council will discuss are likely to be the merits and demerits of new applicants; perhaps the Russian ambassador will be expected to volunteer or even to be asked for his country's opinion on the subject.

What needs to concern us is the fact that under the agreement Russia becomes, at least from its own perspective, a largely self-appointed entity in the very midst of NATO. Its role and its prerogatives may be ill defined and fuzzy, but its presence is unmistakable. This might not be such a bad thing if it happened by the virtue of Russia's efforts to adopt the vision of Atlanticism that we share rather than by the virtue of size combined with weakness used as a negotiating tool. The fate of the bargain largely hinges on whether NATO will be more successful in exporting its strengths to Russia than Russia in exporting its weaknesses outside, something it has done successfully throughout its history, with disastrous consequences to itself and others. This latter possibility would lead to the gradual deconstruction of NATO, something we should not be prepared to accept.

PETER W. RODMAN (United States): We have given Russia an assurance that we will not move nuclear weapons eastward. We have given

Russia an assurance that no substantial conventional combat forces will be moved eastward. There have been significant changes in NATO's doctrine going back to the Bush administration designed to reassure Russia that NATO has only defensive and no offensive intentions.

We have reopened the Conventional Forces in Europe Treaty because of various things the Russians have wanted. We have reopened the START process as well to answer some Russian concerns and to reassure them that the strategic balance will not be threatening to them. President Clinton came close to inviting them into the G7. In the next economic summit, they will have nearly equal status. The Russians continue to receive economic assistance from international financial institutions. Two-thirds of American troops are already out of Europe. In total, NATO has reduced its troop level by 725,000. We have engaged in cooperation with the Russians in peacekeeping in Bosnia. We have offered them a procedure for future collaboration should any similar contingencies arise through the mechanism of the combined joint task forces within NATO. We have set up various cooperative institutions with the Russians including the Partnership for Peace. Now there is something called the Euro-Atlantic Partnership Council that gives Russia a voice and, arguably, a possible veto over the North Atlantic Council.

I hope we pay attention to what we get in return, and I would like to see, as part of this deal, an understanding from the Russians that this is the end of the dispute with them over NATO enlargement. Our interpretation ought to be that this is the end of the matter; that we have made qualitative changes to the alliance to demonstrate beyond any shadow of doubt that it cannot possibly threaten them.

JOSEF JOFFE (Germany): The question today is, Did we concede too much to the Russians? Now, on balance my hesitant answer is no, not really. We did not bribe too much. First, the Paris agreement is a declaration rather than a treaty under international law, and it will not be ratified by national parliaments. This is what the Russians wanted, but that is not what they got.

Second, the three key concessions made to Moscow are at least suitably hedged. There will be no nuclear weapons in the new territories. We did not renounce nuclear weapons under international law but issued a statement that we saw no reason to put them in the new territories, nor do we intend to do so. The same applies to troops: we will not base NATO troops on the territory of the new members, but there is no binding obligation. Russia's relationship with NATO is the most crucial item of all, and we still need to learn a great deal more about how that will play itself out.

Why is the third item, the relationship, so crucial? Russian strategy in this game was to get what Stalin, Brezhnev, and the other leaders had wanted since the early 1950s: to abolish NATO in favor of a collective security system, a perfectly logical strategy from the Russian point of view. Over time, the terms kept changing, but the thrust never did. It was called comprehensive security, overarching security, all European security, or, in the final chapter, the common European house. The basic idea was, though, that Russia would be inside the tent with a decisive role in the strategic arrangements of Europe. If that had worked, the alliance would have dissolved in favor of one of those hapless collective security systems enshrined in the League of Nations and the United Nations.

What is collective security? Everybody is in, including the potential enemy, and everybody swears to uphold the peace against everybody else. We all know what happened to collective security in the 1930s when Italy and Japan took the road to war. Collective security does not work for the same reason that socialism does not work. When everybody is responsible, nobody is responsible. And when everything is owned by everybody, nobody owns anything.

I believe that the West has finessed this thing as well as could have been hoped for. Russia will not be inside NATO, but there will be this Russian-NATO council. The council will comprise a chairman (the secretary general), a vice chairman (a Russian), and a third member. What will they decide? That, of course, will bear the most careful watching. In the worst kind of situation, let us imagine, for example, that Poland feels threatened by Russian forward deployments in Belarus as a result of the recently reestablished union and by an increased state of readiness in the Kaliningrad District, with some 500,000 Russian troops. Feeling this pressure, Poland goes to the NATO Council and says, "You've got to help us." What happens now? Will NATO then sit down with Russia to decide what NATO is to do about Russia? Nothing could be more absurd, and nothing would destroy NATO more quickly than the kind of situation I have just described. All in all, I think the West has done as well as one could have hoped, given that it had previously extended such a golden opportunity for the Russians to pressure and blackmail by weakness.

RICHARD N. PERLE (United States): What struck me about the NATO-Russia Founding Act is that if one were handed this document, without the benefit of any associated information, and asked to say who won the war at the conclusion of which this document was signed, one might well conclude that it had not been won by NATO.

The document has all the markings of one drafted, in substantial part, by Mr. Primakov and his staff. This could be a cold war document slightly revised. In fact, it probably is a cold war document slightly revised. The Founding Act, which is a complicated document, must not become the basis on which NATO's enemies continue their long-established purpose of destroying NATO. It could become that, depending on how it is implemented and depending, most important, on how it is understood. If it is understood as a product that has achieved a consensus leading to cooperation between NATO and Russia, then I think we are in for very serious trouble. If it is the kind of document that we would expect Foreign Minister Primakov to extract in negotiations with the secretary general of NATO—one with the aim of moving the goal posts a bit but that is not intended to achieve a real reconciliation—then I think we can live with even its most troublesome passages because we will be alert to the reality.

Some have urged us to be realistic. The reality is that the authors of the Russian side of this document do not mean NATO well and that they have looked for opportunities to assert Russian positions in a way that is not intended to strengthen NATO but rather to weaken and ultimately destroy it. Nevertheless, there is a range of opinion on the Russian side, and within the context of this document and the obligations that it commits us to, we can be clever and effective or we can be naive and ineffective. I hope we will implement it with an awareness that it is a dangerous document unless we understand its implications.

It would be entirely appropriate, in my view, for the U.S. Senate, in acting, as it must, to ratify a treaty accepting the accession of new members, to examine the Founding Act very carefully, to interpret its meaning, and to guide and instruct the administration with respect to its implementation. That seems to me a responsible role under our constitutional system for the Senate of the United States to play. And so without actually ratifying the Founding Act, a great deal could be done to shape the way in which the United States implements it. I hope the Senate does that and, in doing so, displays real sensitivity toward the potential of the Founding Act to be implemented on the Russian side in the spirit of the continuing confrontation against NATO, rather than in the spirit of cooperation.

BRIAN BEEDHAM (United Kingdom): From the parts of the document that I have seen, NATO appears to be saying that we see no reason to do certain things, but if something were to happen, we could change our minds. There is nothing binding in this document, no absolute promise. I can live with that terminology as far as I understand it.

RAINER SCHUWIRTH (Germany): In my view, during the development of this document NATO showed its strength and passed its test. The document clearly preserves NATO's right to decide on its own. It preserves future members' full rights and obligations and can provide a starting point for developing closer relations with Russia. I am hopeful because this is a process that should be promoted without losing sight of realities; let us not be opposed from the outset.

FRANK J. GAFFNEY, JR. (United States): Just as the United States required, as a condition of the ratification of the CFE flank agreement, that the ABM treaty multilateralization agreement be submitted for the Senate's advice and consent, so should Congress insist on approval of the Founding Act. The concern in the multilateralization context reflected the fear that these documents could, in effect, facilitate and even legitimate Russian or Soviet imperialism. The Founding Act is bound to create a certain dynamic tension if it is ratified as a formal treaty by one party but treated as just an executive agreement by the rest of us.

This document has the potential to destroy the alliance. Perhaps it can be fixed. A German Bundestag member suggested that that could be done as part of a ratification debate. If the Founding Act cannot be fixed, though, I hope that thirty-four senators, at minimum, will reject it. That would not stop NATO enlargement, but it might stop the process of unraveling this alliance, which I think is entirely possible if this agreement goes forward as it is written.

CHARLES GATI (United States): The Founding Act is an ambiguous document, as these documents all tend to be. But to keep NATO strong and united, we must bear in mind West European views. From what I know of the early days of this process, not a single West European government would have been supportive of NATO enlargement had it not been for the U.S. administration's outreach to Russia. In an objective look at this document, I see some things I do not like. I do not like the idea, for example, that it precedes enlargement. I believe that gives the wrong signal, namely, that the Russia-first mentality of some people in the U.S. administration is still with us.

Russia

CHRISTOPHER COX (United States): The collapse of the Soviet Empire, which was, in one sense, a success in and of itself, presented this generation with its greatest opportunity to advance Western ideals of democracy, free enterprise, the rule of law, human rights, and the concomitant stability that the practical implementation of these ideals brings. From a security standpoint, the collapse of the empire offered the opportunity to consolidate the gains of our victory in the cold war by readmitting the formerly captive nations into a free Europe, and it offered the free world a major long-term project, which was the construction, as if from scratch, of a free, democratic, and peaceful Russia.

But for the entirety of the current decade, America, as the leader of the free world, has joined with our European friends in wasting this opportunity. America clung to Gorbachev to the bitter end. We were the next-to-the-last country on earth to recognize the renewed independence of Lithuania, just one step ahead of Cuba, which held out one day longer. We did little to promote the rule of law and free enterprise in the former Soviet Union, but we did, in fact, extend aid to government structures at the very time that their dissolution would have promoted our objectives.

And for all the intervening years, through such temporizing constructs as the Partnership for Peace, we have refused to accept the newly freed nations of Europe into NATO. At the March 1997 Helsinki Summit and in the NATO-Russian Founding Act, we formally admitted Russia into the NATO decision-making process, well ahead of the admission of even the first candidates of nations for NATO admission.

The April 2, 1997, reunion of Russia and Belarus puts us in an odd position. We must now observe that Russia will expand before NATO does. That expansion will bring Russian power, after an absence of only six years, back to the eastern borders of Poland and the Baltic States. Russia's troops in Kaliningrad are not restricted by the limitations of the CFE Flank Agreement. Russia's armed forces have seized parts of Moldova, which, of course, is physically separate from Russia but is contiguous to NATO candidate Romania.

Russia has repeatedly intervened to destabilize and subvert the strategic republics of Georgia and Azerbaijan, the latter of which has newly found and exceptionally important gas reserves, whose transit routes westward Moscow seeks to control. Russia has got forces stationed in Armenia and Tajikistan. I note these facts not to suggest that Russia is our enemy but that Russia is not so concerned about our expansion of NATO because NATO poses a threat to Russia but because NATO poses a threat to Russia's own expansionist aims.

Russia is also spending some money on its defense in interesting ways, not on its current military budget but on research and development. Russia, like China, is, as it were, skipping competitiveness in the current year or in the next few years and putting its chips on R&D down the line. Our analysts are estimating that the Russian R&D budget has risen from $2 billion in 1994 to $13 billion in 1997 and that its total other defense spending, by way of comparison, is $19 billion. The Russians appear to be neglecting their current military assets to concentrate on developing next-generation technologies. They have, it is recently reported, developed new chemical and bacteriological weapons, including a new strain of anthrax, which antibiotics cannot counteract.

Again, all of this is not to suggest that Russia is now our principal military enemy but to suggest something else: in this case, that we would be wise not to assume that it is incapable of becoming that in the longer term.

MAX M. KAMPELMAN (United States): NATO has helped us end this century with a splendid victory in the cold war. The losers of that war were the Soviet Union and the Communist ideology it represented. Russia and the Russian people did not lose. History should judge the Russian people to have been the first victims of Soviet totalitarianism and thus winners with its demise.

Although I make no prediction about what may happen, I certainly hope that the Russian people and their government will be part of an evolving relationship with the rest of us that would permit them to apply for membership in our organization. That would not trouble me in the least.

RADEK SIKORSKI (Poland): The problem, as far as we are concerned, with the NATO-Russia charter is that Russia has a knack of corrupting every organization that it is a member of. It is the old story of allowing Uncle Joe to be a member of the club and imagine that he will behave like a gentleman. That was the logic of Yalta.

More recently, Russia entered the Council of Europe, despite failing to honor the human rights provision of the organization. It is a member of the OSCE but does not respect its principles—for example, look at Chechnya. The same goes for the International Monetary Fund, where rules are routinely broken and, nevertheless, the money keeps flowing. I am just sounding a word of warning. The same may happen with NATO. We are particularly worried in Central Europe about this because we are weak and vulnerable, and we have the experience of 1938 and 1939, where solemnly signed treaties were not honored.

JOSEF JOFFE (Germany): Russia's position today in the European scheme of things is comparable to the position of France after 1815: the great strategic and ideological threat of yesterday, a threat that had to be both contained and resocialized, both watched and rehabilitated. Russia, with its eleven time zones, is, in a way, too big for Europe. It is a country without really fixed borders. Indeed, it is a country that is trying, in various ways, to reconstitute its former empire, be it by suasion, pressure, or outright war, as in Chechnya.

Although I am a good deal more optimistic today than I was five years ago about the prospects for democracy and market economics in Russia, I think it is too early to bet the store on the success of the experiment. Why? Because there is nothing in Russian history that contains the roots of a liberal, democratic, and capitalist tradition. The comparison of Germany does not quite work because in 1945 Germany could look back at very strong liberal market economics and democratic traditions in the past. Russia, by dint of history and its potential power, is still, as it has to be, on probation as was post-Hitler Germany for almost forty years. But we could not quite bring ourselves to say that Russia is the residual threat against which a residual insurance system has to remain in place.

What we did, understandably, was to invite Russia back into the community of the responsible great powers. We had to do that and legitimately so—in fact, as did Europe with France after the Congress of Vienna and as Europe did not do with Weimar Germany, with disastrous consequences.

We have kept repeating the mantra that Russia has no veto. But, in fact, it has. Why otherwise would we have spent the past two years offering ever larger pieces of candy to Boris Yeltsin? I do not blame the Russians for this because the Russians behaved perfectly rationally when handed this opportunity. Of course they kept raising the price, and they did so with a well-known strategy in this test-of-power-and-will game. When someone threatens to go crazy, to lose control over himself, others will do his bidding. The Russians kept telling us that enlargement would have all kinds of catastrophic consequences: nationalism, chauvinism, the victory of the Fascists on the Left and on the Right, the return to militarism and imperialism. And, as a result, we did what was expected. We kept propitiating them, just as we do with our children when we go on a vacation.

BRIAN BEEDHAM (United Kingdom): In the next ten or fifteen years, Russia is likely to remain a fairly marginal power that nine chances out of ten presents no serious threat to an alliance of the democracies. This is a country whose GDP is now falling for its sixth or seventh year in succession.

This is a country whose defunct nuclear submarines litter the shores near the northern coasts of Norway. This is a country where the life expectancy of a sixteen-year-old male is now less than it was a hundred years ago. This is a country in profound economic straits; it will take a long time to regain economic, psychological, and military strength and in the interval will very much need good relations with the West.

When it does emerge, Russia will have more cause than almost any other country in the world to feel concern about the new power of China to its east. Russia has longer and more vulnerable borders than any other country with this potential new Chinese power. It has a set of old quarrels that have been patched up only in a formal sense. The eastern section of Russia is astonishingly empty and is therefore vulnerable. Russia will, in my view, turn toward the West for reassurance and probably assistance in this second period.

There is a one in ten chance that Russia will not behave rationally during this period, as I have been assuming it will. There is a slim chance that a nationalist reassertion will bring to power a government pursuing irrational foreign policies. But that scenario, on the whole, seems fairly unlikely. The history of Russia produces periods of great grumpiness, but it seldom becomes hysterical. Look at what happened to Mr. Zhirinovsky, this star of the irrational nationalist Right, who fell to Earth quite rapidly.

Do we take Russia into NATO? No, not on the argument that Russia is too big a country to be left out. After all, we have one quite large country in NATO already. But also, and more important, Russia's interests, particularly in its relations with the Muslim world to its south, are by no means necessarily the same as ours. It makes sense to think of Russia as an entity separate from the alliance of the democracies for quite a long time ahead. But there is no reason that it cannot be a reasonably good neighbor. I am happy that the recent arrangements bring Russia into a position where it can watch NATO policy making. I think that watching what NATO is thinking about will reassure the Russians that we are not aiming NATO at them. It also gives us the chance to invite them to offer reciprocity. If they are watching how our foreign and military policies are being discussed and taking shape, we can claim the same right from them. We have not yet claimed that reciprocity, and I am sorry that this opportunity was not included in the recent agreement—but we should ask for it.

Russia should, of course, have no veto, and I think it is now clear that it will not have one. Our aim is a Russia that gradually accepts and understands that NATO is not aimed at it and that can, in certain circumstances, cooperate with NATO but will remain, as it were, just on the other side of a boundary. But as we Europeans know and Americans remind us, good fences make good neighbors; that is the approach I want to take to Russia.

WILLIAM E. ODOM (United States): Referring to Russia in the singular troubles me; in doing so, we may overlook some important realities about Russia. In Moscow, for example, dealing with the Russian government is difficult. Are we dealing with the presidency? Are we dealing with the prime minister? Are we dealing with the Security Council? Are we dealing with the Duma? On the outside and virtually nonexistent chance that we could get all of them to take one position on anything for more than a day, we would be agreeing only with Moscow, not with Russia, because Russia's control over a large part of the rest of the country is tenuous at best.

I disagree with those who believe it is possible to reassure that kind of government, and to think it is possible is a misperception of reality. The concerns about "committing the error of Weimar" or worries that we have made Russia "a pariah state" and failed to give it a Marshall Plan seem misplaced to me. I see no record of Russia's introducing liberal reforms that have not come from heavy outside pressures and crises within. Concessions are not rewarded. The liberal reforms after the loss of the Crimean War, the reforms of Alexander the Great, those came on the heels of defeat. The October Manifesto and the Duma in 1906 followed the defeat in the Russo-Japanese War. The provisional government came upon exhaustion in World War I, and Gorbachev came in the wake of losing the qualitative arms race and a defeat in Afghanistan. Those examples tell us that we cannot depend on concessionary activities to bring in reform. I do not favor refusing to consider any concessions whatsoever or seeking noncooperative relations, but we are not far enough along with the transition of Russia from its old condition to a new condition to treat it as we would some other European country.

There is not even a consensus among Russians over what constitutes Russia, which is a problem. We are dealing with a highly irrational Russian policy because nobody is in charge, and we can expect that Russia will behave erratically. I do not see its military strength as the major problem with regard to Europe but rather its trouble-making diplomacy. Speaking from an American point of view, I think London, Paris, Bonn, and, in the future, Berlin will be susceptible to Russian mischief and will be played off against one another by an extremely weak Russia.

Although I do not think we should be upset over the Founding Act, any such paper in the hands of a clever, devious, trouble-making diplomat could be the basis for considerable difficulties unless the West is united and determined not to let matters develop beyond a nettlesome irritation.

DOREL SANDOR (Romania): A common illusion about Russia is that its decline as a superpower will continue during the next ten years. Another is that, as a consequence of this decline, we will have democratization and marketization of the Russian economy. It should now be obvious that Russia is prepared to restore its empire and has already started to do so. Progress has so far not been very significant. But after Belarus, Ukraine may become the next target.

That is the reason that Romania was so flexible in negotiating a treaty with Ukraine. We understand better than most that the stability and consolidation of Ukraine as an independent state are one of the best guarantees for security in relations with Russia. We should thus pay attention to this possible evolution of Russia. Our second concern is that Russia may seek to restore a strategic alliance with China. In some respects, they are very similar post-Communist societies. Our third concern has to do with the leadership in Russia, which could revert to an extreme nationalist or authoritarian style. Russia without Yeltsin will be highly unpredictable, but Western leaders behave as if Yeltsin will be the leader of Russia forever.

MR. SIKORSKI: I wanted to respond to the point that we are dealing with some rather nasty people out there. I remember watching in stupefaction a press conference given jointly by Al Gore and Prime Minister Chernomyrdin, in which the vice president was stressing over and over again how much trust, friendship, and respect he felt for his Russian counterpart. We now know that Mr. Chernomyrdin owns 1 percent of Russian gas; in other words, he has stolen $5–8 billion from his own people.

Those gangsters running the country are trying to come across as legitimate. Foreigners arriving in the United States must fill out a landing card and state whether they were associated with the persecutions by Nazi Germany or its allies during the war. Foreigners do not have to state whether they had part in Communist persecutions. A gulag commander from ten years ago is perfectly welcome in the United States. From our Central European perspective, this whole business of wooing Russia sounds to us like saying, "We'll defend you against Russia, perhaps, provided Russia agrees." This is not very credible, which is why we are afraid of the repetition of the British guarantee to Poland of 1939. We will credit NATO only when those American and NATO troops arrive in our countries and the infrastructure is extended.

Finally, I agree with Mr. Sandor that Russia will not be weak forever. That country has enormous resources, and it will eventually learn that it can get American companies to mine all of those diamonds and all of that oil for them and just rake in the profits. Those profits will be put to uses that are not necessarily congenial to us.

MARTIN SIEFF (United States): We see, as happened in Weimar Germany, a country that is clearly unstable, clearly dissatisfied with its place in the world and that contains extremist elements. Peace-loving Weimar, too, we now know, was secretly reequipping and planning ahead for the reassessment that even then it was sure would come. I would follow the Weimar analogy another two steps. First, several have noted that we should not repeat the Weimar mistake by making Russia an outcast from the West. In fact, though, Weimar was not an outcast from the West. It signed the Locarno Treaty in 1925. It agreed to outlaw war forever in the bizarre Kellogg Abolition of War Pact, for which Secretary of State Kellogg won the Nobel Peace Prize. Therefore, Weimar is a parallel to Russia today.

DAN QUAYLE (United States): We need to be realistic about the future of Russia. Yeltsin is probably as good as we are going to get, whoever his successor may be. That successor could be someone with a more nationalistic viewpoint, someone who, perhaps, comes from the military or has a military background, someone who knows how to reenergize the country to seek to recover its superpower status by military force. It is important to be realistic, especially given the fact that Russia still has 50,000 weapons of mass destruction, including the intercontinental ballistic missiles that can hit any city in the world in a matter of minutes.

What is the future of Russia? We will wait and see—we can work with the Russians, we can be a friendly partner, and we can try to advance democratic values and economic freedom in that country. But the challenge is steep.

Ukraine and the Baltics

CHRISTOPHER COX (United States): What about admitting the Baltic countries into NATO or admitting Romania or others? It has been said that the Baltic States make particularly poor candidates for NATO admission because of their inability to contribute militarily to the alliance. But if need for participation in the defensive arrangement of NATO were the criterion instead, then these would be the best choices to join NATO, and, in fact, they would remind us of the way in which the alliance welcomed Iceland and Luxembourg as charter members. Surely that same argument would apply to the Baltics. NATO even vowed to go to war if the Soviet Union attacked militarily indefensible West Berlin. That was as key an element of the original organization as one can imagine.

ANTON D. BUTEYKO (Ukraine): The independence of Ukraine is critically important to the process of cooperation and integration of the territory of the former Soviet Union. NATO enlargement should be considered the most powerful stimulus in support of the democratic trend within the Russian society and within the former Soviet Union. Without exaggerating the role of Ukraine, I believe that an independent Ukraine would bury the hopes of those who would like to revitalize the old empire. But if Ukraine is under the influence of Russia and loses its independence, then the strength of those forces within Russian society will be increased and, therefore, undemocratic tendencies will prevail.

OJARS KALNINS (Latvia): Based on what has happened in relation to the Baltics recently, we need to add one more item to the list of concessions or bribes that have been given to the Russians: essentially a veto on Baltic membership in NATO. Officially, of course, there is no such veto, but everyone knows that it is totally out of the question that the Baltics could eventually become members. Unfortunately, we are seen as high-risk candidates. As many have said, those who need membership most are those least likely to get it. The suggestion seems to be that Russian aggression against the Baltic States is more likely if they are in NATO than outside it and that joining NATO would, in fact, prompt an aggressive reaction. That seems unlikely, given that such an act would also prompt an Article V reaction.

Ten years have passed since the Baltic States gained independence, and the Baltics are alive and well. We believe that Russia has finally learned to live with our independence. Why couldn't Russia eventually learn to live with the Baltic States as part of NATO?

MR. BUTEYKO: Our new constitution does not prohibit Ukraine from joining the alliance. And in the past there was a document proclaiming our nonbloc status, but the constitution was passed after that document. In addition, neutrality is a useful institution, and under the Paris declaration every country has the right to choose the methods to ensure its security. These institutions can be applicable, but it is up to the country itself to choose whether to protect its security individually or within the alliance.

MARTIN SIEFF (United States): Ukraine today is indeed crucial. It is where Austria was in 1938. It is the canary in the coal mine. The first signs of an aggressive Russia will be moves to destabilize Ukraine.

DAN QUAYLE (United States): We ought to explore the possibility of having a free trade agreement with Europe. And if Europe does not want go along with it, we can start specifically with the Baltic nations. Maybe we should just leapfrog Europe.

Bosnia

MAX M. KAMPELMAN (United States): Never since World War II has the European and American head been buried in the ground so irresponsibly. This is not the time to review that tragic history, which will continue to plague us as we pay a price for our shortsighted neglect. My concern is with the blatant decision by the United States and our European NATO partners to undermine the authority and capacity of the International War Crimes Tribunal to function. Of seventy-five persons indicted, eight relatively small criminal actors are in custody, while the others roam free.

The police force of every community in the world understands that, by definition, there is some danger in the process of arresting a suspected criminal. But the task of arresting criminals is indispensable for a civilized community to exist in safety. Yet the president of the United States announces, and Europe agrees, that our troops will not arrest those indicted by the International War Crimes Tribunal, killers whose whereabouts are fully known, because that would put our soldiers in harm's way. The result is that those responsible for fomenting and executing brutality go brazenly free, continuing to exercise power and intimidate refugees, while the court created by the United Nations Security Council to impose justice and deter future butchery is left in limbo, the object of mockery by those who escape its jurisdiction. This is no way to enter the twenty-first century, if we expect in that century to achieve a more stable and just international community.

RICHARD N. PERLE (United States): There was a moment not long ago when we all understood that a failure in Bosnia represented a fundamental threat to the integrity of the alliance. That recognition will come back to us as we face the withdrawal of NATO forces from

Bosnia and as we are again forced to ask ourselves whether NATO can honorably and safely withdraw from Bosnia without destroying its integrity.

We must now face the fact that the political dimensions of Dayton—the promises to facilitate the return of refugees and to apprehend war criminals—have not been dealt with effectively and will not be dealt with effectively until there is a change of policy with respect to the Dayton accords.

Turkey

RICHARD BURT (United States): In some respects, in a post–cold war period Turkey has the same role or position within the alliance that Germany had during the cold war. Turkey has become *the* front-line state. The problems that Turkey faces, including internal cohesion, the threat of radical fundamentalism, and the threat of terrorism, highlight the problems that our societies will face in the future.

Both American and European policies toward Turkey in the past five years have been nothing short of shameful. Turkey has tried every means that it can to move closer to Europe, to be a member of the union—and if not a member of the union, to maintain its Western orientation and its secular traditions. It has been soundly rebuffed by the European Union and many serious politicians in the union. At the same time, the United States has isolated Turkey. America has conducted a virtual arms embargo on Turkey at a time when Turkey is facing many serious threats on its borders. The results of this policy have been to create a very delicate and unstable situation within Turkey itself. Turkish institutions are under attack, while Turkey simultaneously faces severe threats on most of its borders.

I propose that one clear mission for the alliance and a new definition of Atlanticism would be for Europeans and Americans alike to shore up their support for Turkey, its democratic institutions, its secular traditions, and its national security.

NORMAN STONE (United Kingdom): Samuel Huntington's book, which suggests that Islamic civilizations are stagnant, dangerous, or dead, is almost a self-fulfilling prophecy. It is the kind of thing that can create a great deal of tension in countries like Turkey. And if Europe does get Turkey wrong, as it is in danger of doing, the consequences could be profound. I am constantly struck by the nitpicking way that the Europeans are handling Turkey—with quotas for exports and all sorts of unsuitable remarks coming from the European Parliament, a parliament that represents countries that within living memory were slaughtering people on a massive scale but that now come along and preach in this irritating way to a country that in 1922 could not make a table and in 1997 can produce a fax machine.

The Turkish growth rate has now overtaken the government statistical machine. It has a growing population, growing perhaps too fast, but it is not at all unthinkable that in thirty years' time Turkey will be sitting on our doorstep, overtaking the fat, rich, contented countries of Western Europe. And some very old men from Bonn might turn up in Constantinople and say, "Please, can we join the Ottoman free trade zone?"

The monotheistic religious background has been very important for Atlantic civilization, for Judaic civilization, and so it is in Islam. One of the big problems in the Western world is social disintegration, something we should all be thinking about more seriously. It is very strange and worrying to see that the more prosperous we get, the more children are born out of wedlock. And the religious sanctions that discourage that from happening are enormously important. So it would be wrong to launch a frontal attack on Islam simply because it believes in things like paternal authority. Why not? We used to believe in that sort of thing ourselves not so long ago.

Islam is not the monolithic entity often assumed. There are different styles of Islam and quite different beliefs. And, of course, there is also a secular tradition in Turkey. This is a state that is separate from the Islamic religion. There are all sorts of difficulties—this is a country with problems. But it is wrong for us in the West to conclude that Islam simply condemns people automatically to a dreadful life. Max Weber's famous book, *Religion and the Rise of Capitalism,* written in 1903, says that the Catholics are all backward. They drink too much. They make too many babies. Catholicism condemns people to eternal poverty. It is slightly shaming for me, as a Protestant Scotsman, to see that Catholic Ireland is becoming richer than we are. And it is simply nonsense to say that a particular religion condemns a person to eternal poverty. A more accurate explanation to account for the decline of the Mediterranean, including the Ottoman Empire, is simply malaria. It accounts more logically for the decline than does the theology of Protestantism.

In conclusion, I think NATO should push the Europeans to be a lot more understanding about what the Turks face in the way of problems and what sort of solutions the Turks can actually bring to us. In the Renaissance period, scholars famously went from Byzantium, which was being beset by the Turks, to northern Italy. Although the matter may be exaggerated in the way these things often are, they contributed a great deal to the Italian Renaissance. Now, we can perform the process in reverse. Europe must be encouraged to take Turkey more seriously than it is now doing. Otherwise, we are going to take this very promising, growing country and turn it into a festering slum on our doorstep.

MARGARITA MATHIOPOULOS (Germany): Another issue is Turkey's EU membership. I believe it is very important that we integrate Turkey into the European Union, but I think we have to be consistent because we cannot work with double standards. If we talk about human rights in China, we have to talk about human rights violations in Turkey as well. But it is a good incentive to tell Turkey to behave like an Atlanticist country and adopt the rules, principles, and values—and then the door will open. This can happen in three to five years. I think it is important that we keep Turkey strong.

Missile Proliferation and Strategic Defense

MARTIN SIEFF (United States): I am reminded of several nostrums. For example, an ounce of prevention is worth a pound of cure. In 1940, after all the diplomatic permutations that had taken place in previous years, all that could have been done and was not done, ultimately, the survival of the West depended on one fact: that the British government, despite its other failings, had under Churchill's proddings developed a handful of squadrons of advanced monoplane fighters with eight Browning machine guns per plane. That was sufficient to stop the Luftwaffe from bombing Britain into submission. The equivalent for us today is the Strategic Defense Initiative (SDI). This is particularly relevant to the uncertainties concerning Russia. Brian Beedham has estimated a one in ten chance of some ugly and unpredictable irrational regime taking over in Moscow sometime after President Yeltsin departs from the scene. We all know that anything can be proved with statistics, but I would like to know what his quantitative analysis was that brought him to a merely 10 percent chance. I think that the figure is unquantifiable. We simply do not know, and that is an extremely alarming situation to be in.

Twice within the past seven years— in August 1991 and October 1993—we now know that the situation was so uncertain to all the players in it that no one knew who or what would emerge as the rulers of Russia. General Odom makes the point that the Russian army is relatively small and weak, a force of only 1.2 million. And we know that it performed miserably in Chechnya. But look how hideously it performed against Finland in 1939 and 1940, not to mention the collapse that followed on the front of Smolensk in Ukraine the following year. And then within a year came the recovery at Stalingrad in the most difficult conditions. As Professor John Erickson of Edinburgh University, one of the world's leading experts on the Soviet military, is fond of saying, it is a mistake to underestimate Russians because they drink a lot. They were drunk when they destroyed Hitler, and they were drunk when they annihilated Napoleon.

Specifically, if one looks at the strategic balance, Russia is a nation with at least 20,000 nuclear weapons, perhaps 27,000. No one knows for sure. Quite probably the Ministry of Defense in Moscow itself can no longer keep track of all its tactical nuclear weapons. We do, however, know this: the Russians have kept their strategic weapons systems at a high level of priority. The catastrophic erosion that has occurred in their conventional military has not affected strategic rocket programs to anything like the same degree. There has been serious deterioration, but that could probably be reversed with a couple of years of serious investment, which could feasibly happen within the next decade.

DAN QUAYLE (United States): The ABM treaty was signed in 1972. It is inoperative. It is simply obsolete. In this situation, we have to look at what is in our best interests and then to go ahead with research to have the scheduled deployment of some ballistic missile defenses in this country. One of today's threats is ballistic missile proliferation. Some twenty-five different non-NATO nations have ballistic missile capability. Why wouldn't we want to defend ourselves against a ballistic missile attack?

Ask the American people what the response to a ballistic missile attack on the United States should be, and they would say to shoot down the missiles. When told that we do not have the capability to shoot them down, they are dumbfounded.

China

DAN QUAYLE (United States): I know the temptations that China presents to the West. I serve on some corporate boards, and they all want to go over there and invest. Fine—let's invest: China has 1.2 billion people and is growing. But let's do it with open eyes.

What is the future of China? I have a vested interest in China because my oldest son, Tucker, who is fluent in Mandarin Chinese, is at Szechuan University teaching English and will be working for an American corporation in China for the next two years. According to him, there is a tremendous change going on in China, but the Chinese do not know which way they are going to go. I urge everyone to read *The Coming Conflict with China* by Richard Bernstein. As he points out, China has stated that it views us as the enemy, and its military strategy reflects that. Why is it that we don't listen to what the Chinese say and what they publish? I do not view China as an enemy. I believe there is an opportunity for it to begin the process of joining the community of democratic nations so that eventually it will come to respect the sovereignty of other nations and individual rights.

Many have suggested an analogy with the Germany of a hundred years ago. How long did it take us to resolve the situation in Germany, and is that situation finally resolved even today? I think so. But how long and how many casualties will there be, and what costs will be incurred before the China question is resolved? We should be realistic about it. Mischief Reef was taken by China several years ago, and Hong Kong was peacefully transferred. Macau Island will go next in 1999, and after that it will be Taiwan. This is no secret. Why is it that we sometimes don't want to believe what people say? Of course, they mean it. This is their strategy. Why do they have a blue-water navy or are in the process of developing a blue-water navy? To take care of the things in inland China? To guard against a potential new Mao-type revolution? I doubt that. Blue-water navies are not used for that.

Under Chinese law, the *Financial Times* will not be able to publish in Hong Kong. Under Chinese law, the Asian edition of the *Wall Street Journal* will not be able to publish in Hong Kong. Will Chinese law be imposed in Hong Kong? I presume so. They have said that it is going to be. Take them at their word.

Appendix A

Agenda for the Congress of Phoenix

May 16–18, 1997

Friday, May 16

Opening Plenary Session

Welcome
Christopher DeMuth

Introductory Remarks
John McCain and
Jon L. Kyl

Keynote Address
Paul Wolfowitz

Night Owl Discussions

Privatization and Deregulation: Favorite Stories
James K. Glassman and Wilfried Prewo, *chairmen*

European Security after Madrid
Paula J. Dobriansky and Gilda Lazăr, *chairmen*

Saturday, May 17

Plenary Session

The Case for New Atlanticism: Four Perspectives
Edward Streator, *chairman*
Christopher Cox, Charles Powell, Max M. Kampelman, and
Michael Zantovsky, *presenters*

Reply
James K. Glassman

BREAKOUT SESSIONS: ATLANTIC ECONOMIC TENSIONS

The West and the Rest: Trade, Sanctions, and Strategic Priorities
James F. Hoge, Jr., *chairman*
Ruprecht Polenz
Claude Barfield

West-West Issues: Currencies, Markets, and Barriers to Prosperity
Allan H. Meltzer, *chairman*
Christian Stoffaes
Gunter Dunkel
David Davis

Facing Asia
George Melloan, *chairman*
D. Howard Pierce
Andreas Krautscheid

LUNCHEON

Welcome
Douglas C. Yearley

Introductory Remarks
Irwin Stelzer

Keynote Address
Margaret Thatcher

BREAKOUT SESSIONS: NATO: NEW THREATS, NEW STRUCTURES

NATO Expansion and the Russians:
Are Bribes Necessary? Which Bribes Are Excessive?
Peter W. Rodman, *chairman*
Brian Beedham
Anton D. Buteyko
Josef Joffe

NATO Expansion: The Costs
Jeremy D. Rosner, *chairman*
Robert Mroziewicz
Réka Szemerkényi

PLENARY SESSION

Deepening the Atlantic Community

Introductory Remarks
William E. Odom

Keynote Address
Jon L. Kyl

Reply
Nuño Aguirre de Cárcer

NIGHT OWL DISCUSSIONS

Bosnia
Richard N. Perle and Christian Schmidt, *chairmen*

The United States, Europe, and Britain's New Government
David Goodhart and John Bolton, *chairmen*

Sunday, May 12

PLENARY SESSION

The Future of Atlanticism

Opening Remarks
Jeffrey Gedmin and John O'Sullivan

Keeping NATO Strong
Richard N. Perle, *chairman*
Rainer Schuwirth, Charles Gati, and Andreas Unterberger, *presenters*

Complements to NATO:
Revitalizing the Old and Creating New Atlantic Ties
Adrian Karatnycky, *chairman*
Thomas J. Duesterberg, Margarita Mathiopoulos, and
Norman Stone, *presenters*

Appendix B

Attendees

Nuño Aguirre de Cárcer
Atlantic Treaty Association

Anne Applebaum
Evening Standard

David Asman
Wall Street Journal

Ronald Asmus
U.S. Department of State

Mira Baratta
National Forum Foundation

Pamela Barbey
Office of U.S. Senator John Kyl

Claude Barfield
American Enterprise Institute

Brian Beedham
The Economist

Steven P. Berchem
American Enterprise Institute

Ryszard Bobrowski
Central Europe Review

John Bolton
American Enterprise Institute

James B. Burnham
Duquesne University School of Business

Richard Burt
IEP Advisors, Inc.

Anton D. Buteyko
Ministry for Foreign Affairs, Ukraine

Chris Butler
Grandfield Public Affairs

Joseph A. Cannon
Geneva Steel

Geoffrey Carlson
The Windsor Foundation

William Cash
British Parliament

Patrick A. Chamorel
Powell, Goldstein, Frazer & Murphy LLP

Steven C. Clemons
Office of Senator Jeff Bingaman

Christopher Cox
U.S. House of Representatives

David Davis
British Parliament

Christopher DeMuth
American Enterprise Institute

James Denton
National Forum Foundation

Paula J. Dobriansky
Hunton & Williams

Valentine Dubrov
Foreign Ministry, Bulgaria

Thomas J. Duesterberg
Hudson Institute

Gunter Dunkel
Norddeutsche Landesbank

Jeanine Esperne
Office of Senator Jon Kyl

Douglas J. Feith
Feith & Zell, P.C.

Julie Finley
U.S. Committee to Expand NATO

John Fox
Open Society Institute

Bruce Friedman
U.S. Department of State

Gerald Frost
The New Atlantic Initiative

John Fund
Wall Street Journal

Frank J. Gaffney, Jr.
Center for Security Policy

Evan G. Galbraith
LVMH Möet Hennessy Louis Vuitton Inc.

Charles Gati
Interinvest

Jeffrey Gedmin
The New Atlantic Initiative

Gary L. Geipel
Hudson Institute

David Gelber
CBS News

David Gerson
American Enterprise Institute

James K. Glassman
American Enterprise Institute

David Goodhart
Prospect

Claus E. J. Gramckow
Friedrich Naumann Foundation

Dorothy Stephens Gray
D.C. Stephens, Ltd.

Stephen J. Hadley
Shea and Gardner

Marshall Freeman Harris
The Balkan Institute

Craig R. Helsing
BMW (US) Holding Corp.

Heather Higgins
Randolph Foundation

James Higgins
Columbia Naples Capital

James F. Hoge, Jr.
Foreign Affairs

Stuart W. Holliday
Dallas Council on World Affairs

Kim R. Holmes
The Heritage Foundation

Bruce P. Jackson
Lockheed Martin Corporation

Josef Joffe
Süddeutsche Zeitung

Ojars Kalnins
Ambassador of Latvia to the United States and Mexico

Max M. Kampelman
Fried, Frank, Harris, Shriver & Jacobson

Adrian Karatnycky
Freedom House

Marlene M. Kaufmann
Commission on Security and Cooperation in Europe

Craig Kennedy
German Marshall Fund

Llewellyn King
White House Weekly

Martin M. Koffel
URS Corporation

Maciej Kozlowski
Ministry of Foreign Affairs, Poland

Andreas Krautschied
German Bundestag

Jon L. Kyl
U.S. Senate

Mark P. Lagon
Office of the Republican Policy Committee U.S. House of Representatives

William P. Laughlin
Saga Corporation

Gilda Lazăr
Romanian Foreign Ministry

Michael A. Ledeen
American Enterprise Institute

Elke Leonhard
German Bundestag

Branislav Lichardus
Ambassador of the Slovak Republic to the United States

Michal Lobkowics
Czech Foreign Relations Committee

Sandy A. Mactaggart
Former Executive, Maclab Enterprises

Margarita Mathiopoulos
Norddeutsche Landesbank

Marek Matraszek
CEC Government Relations

John McCain
U.S. Senate

Anna M. McCollister
Radio Free Europe/Radio Liberty

George Melloan
Wall Street Journal

Allan H. Meltzer
Carnegie Mellon University

Jürgen Miele
Hanns Seidel Foundation

Robert Mroziewicz
Ministry of ForeignAffairs, Poland

Joshua Muravchik
American Enterprise Institute

William E. Odom
Hudson Institute

Daniel Oliver
Preferred Health Systems

Louise V. Oliver
The New Atlantic Initiative

Paul F. Oreffice
Dow Chemical Company (Retired)

John O'Sullivan
National Review

George Palmer
Boeing North America, Inc.

Solomon Pasi
Atlantic Club of Bulgaria

Valois Pavlovskis
Baltic American Freedom League

Richard N. Perle
American Enterprise Institute

D. Howard Pierce
The Americas Region, ABB Inc.

Ruprecht Polenz
German Bundestag

Mark Pomar
International Research and Exchanges Board

Charles Powell
Jardine Matheson Holdings

Wilfried Prewo
Hannover Chamber of Industry & Commerce

Dan Quayle
Former Vice President, United States

David B. Rivkin, Jr.
Hunton & Williams

Peter Robinson
Atlantic Council of the United Kingdom

Peter W. Rodman
Nixon Center for Peace and Freedom

Jeremy D. Rosner
Special Adviser to the U.S. President and Secretary of State for NATO Enlargement Ratification

Allen H. Roth
RSL Management

Muhamed Sacirbey
Bosnian Ambassador and Permanent Representative to the United Nations

Ferdinando Salleo
Italian Ambassador to the United States

Matt Salmon
U.S. House of Representatives

Dorel Sandor
Center for Political Studies and Comparative Analysis

Christian Schmidt
German Bundestag

Rainer Schuwirth
Federal Ministry of Defense, Germany

Martin Sieff
Washington Times

Radek Sikorski
National Review

Geoffrey Smith
American Enterprise Institute

W. Richard Smyser
The German Economy

Irwin Stelzer
American Enterprise Institute

Christian Stoffaes
Audit and Control, Electricité de France

Kalev Stoicescu
Estonian Ambassador to the United States

Norman Stone
Worcester College, Oxford

Edward Streator
The New Atlantic Initiative

Johnathan Sunley
Windsor Group (Budapest)

Réka A. Szemerkényi
Hungarian Institute of International Affairs

Margaret Thatcher
Former Prime Minister, United Kingdom

K. Erik Tygesen
Danish Ambassador to the United States

Andreas Unterberger
Die Presse

Arno P. Visser
Office of the Vice Prime Minister
The Netherlands

Allen Wallis
American Enterprise Institute

W. Bruce Weinrod
Allen and Harold

Alan Lee Williams
Atlantic Council of the United Kingdom

Catherine Windels
Pfizer Inc.

Paul Wolfowitz
The Nitze School of International Study
Johns Hopkins University

David Wurmser
American Enterprise Institute

Douglas C. Yearley
Phelps Dodge Corporation

Michael Zantovsky
Czech Parliament

Wolf-Dieter Zumpfort
Friedrich Naumann Foundation

Miomir Zuzul
Croatian Ambassador to the United States

Editor and Contributors

GERALD FROST, editor of this volume, is the research director of the New Atlantic Initiative. An author and journalist, he has written extensively on political issues in the United States and Britain. He was director of the Centre for Policy Studies in London and founder-director of the Institute for European Defence and Strategic Studies.

MIRA BARATTA has served since February 1997 as a senior consultant at the National Forum Foundation. From 1989 to 1996, she was a legislative assistant for arms control and foreign policy to Senator Bob Dole. Ms. Baratta was Senator Dole's adviser on foreign and defense policy during the 1996 presidential campaign. She also served in the Reagan administration as the deputy director for congressional affairs at the U.S. Arms Control and Disarmament Agency.

BRIAN BEEDHAM is an associate editor of the *Economist* and a columnist for the *International Herald Tribune*. He was the *Economist's* foreign editor from 1964 until 1989. He read classics at Oxford and served in the Royal Artillery.

RICHARD BURT is the chairman of IEP Advisors, Inc., a Washington-based financial and business advisory services firm. He was a partner with McKinsey & Company, specializing in international business strategy and telecommunications. Mr. Burt was the U.S. chief negotiator in the Strategic Arms Reduction Talks (START) with the former Soviet Union. He also served as the U.S. ambassador to the Federal Republic of Germany from 1985 to 1989.

ANTON D. BUTEYKO is first deputy foreign minister of Ukraine. He is also a member of Ukraine's parliament, the head of the subcommission on currency regulation of the parliament, and a member of the Ukraine Constitutional Commission. Mr. Buteyko was foreign policy adviser to the president and chief of the legal department of the Ministry of Foreign Affairs.

CHRISTOPHER COX was elected to the U.S. House of Representatives in 1988, representing the forty-seventh district of California. He is the chairman of the House Republican Policy Committee, a member of the House Leadership Steering Committee, the vice chairman of the Subcommittee on Oversight and Investigations, a member of the Subcommittee on Telecommunications, Trade, and Consumer Protection, and the vice chairman of the Committee on Government Reform and Oversight. Prior to his service in Congress, Mr. Cox was a senior associate counsel to the president from 1986 to 1988.

CHRISTOPHER DEMUTH is a cochairman of the New Atlantic Initiative and has been the president of the American Enterprise Institute since 1986. He was previously the managing director of Lexecon Inc., administrator for regulatory affairs at the U.S. Office of Management and Budget, executive director of the Task Force on Regulatory Relief in the Reagan administration, lecturer and director of regulatory studies at Harvard's Kennedy School of Government, and an attorney with the Consolidated Rail Corporation and the law firm of Sidley & Austin. Mr. DeMuth is chairman of two family businesses. His articles on government regulation and other subjects have appeared in *The Public Interest*, the *Harvard Law Review*, the *Yale Journal of Regulation*, the *Wall Street Journal*, and elsewhere.

THOMAS J. DUESTERBERG is a senior fellow and the director of the Washington Office of the Hudson Institute. His research interests include trade policy, strategies for technology development, and regulation of emerging technologies. Before joining the Hudson Institute, he was chief of staff to Congressman Christopher Cox and to Senator Dan Quayle. Mr. Duesterberg also served as assistant secretary of commerce for international economic policy.

DOUGLAS J. FEITH is a founding member of the law firm of Feith & Zell, P.C. He has served as deputy assistant secretary of defense for negotiations policy, special counsel to the assistant secretary of defense for international security policy, and as a Middle East specialist at the National Security Council.

FRANK J. GAFFNEY, JR., is the founder and director of the Center for Security Policy. He is also a founder and coordinator of the Coalition to Defend America, and he is the host and executive editor of the broadcast *The World This Week*. Mr. Gaffney is a columnist for the *Washington Times* and is a monthly contributor to *Defense News*. He was assistant secretary of defense for international security policy in the Reagan administration.

CHARLES GATI is senior vice president at Interinvest, a global money management firm. He is also a fellow at the Foreign Policy Institute of the Johns Hopkins University's Paul H. Nitze School of Advanced International Studies in Washington. Mr. Gati was senior adviser on European and Russian affairs at the Department of State's Policy Planning Staff from 1993 to 1994. He was professor of political science at Union College in Schenectady, New York. He also taught at Columbia University and was a visiting professor at Yale, Georgetown, and Kansas Universities. Mr. Gati is the author of *The Bloc That Failed* and *Hungary and the Soviet Bloc*, as well as several other books and articles.

BRUCE P. JACKSON is the director of planning and analysis on the corporate staff of Lockheed Martin Corporation and the president of the U.S. Committee to Expand NATO, a nonprofit organization formed to promote the expansion of the NATO alliance. He was a military intelligence officer in the United States Army from 1979 to 1990 and served in the office of the secretary of defense from 1986 to 1990. In 1993 Mr. Jackson joined Martin Marietta Corporation, which later merged with Lockheed Corporation. He is a member of the Council on Foreign Relations and the International Institute for Strategic Studies in London and is on the board of advisers of the Center for Security Policy.

JOSEF JOFFE is a columnist and the editorial page editor at the *Süddeutsche Zeitung*. He is a contributing editor at *U.S. News and World*

Report and *Time*. His articles and opinion pieces appear widely in the American and European press. Mr. Joffe is also a talk show host on German television and a frequent commentator on American and European radio and television. He teaches at the University of Munich and the Salzburg Seminar.

OJARS KALNINS is the Republic of Latvia's ambassador to the United States of America and Mexico. From 1985 to 1990 he was the director of public relations for the American Latvian Association in Washington, D.C., the largest Latvian organization outside of Latvia. In 1991, he was appointed minister counselor and deputy chief of mission for the Latvian Embassy in Washington and deputy permanent representative of the Latvian Mission to the United Nations. He is currently Latvia's permanent observer to the Organization of American States. He has written articles on Latvian and Baltic topics for the *Wall Street Journal*, the *American Spectator*, and the *Chicago Tribune*.

MAX M. KAMPELMAN is an attorney with the law firm of Fried, Frank, Harris, Shriver & Jacobson in Washington, D.C. He was counselor of the U.S. Department of State and ambassador and head of the U.S. Delegation to the Negotiations with the Soviet Union on Nuclear and Space Arms in Geneva. Mr. Kampelman is now the chairman of the American Academy of Diplomacy, chairman of Georgetown University's Institute for the Study of Diplomacy, and, by presidential appointment, vice chairman of the U.S. Institute of Peace.

MACIEJ KOZLOWSKI was nominated head of Poland's American Department on December 19, 1994. From 1990 to 1993, he was the deputy chief of mission at the Polish Embassy in Washington, D.C., and from 1993 to 1994, he was the Polish chargé d'affaires in Washington. Mr. Kozlowski was a member of the editorial board of *Tygodnik Powszechny*, a prominent Catholic weekly newspaper published in Krakow. He was a Fulbright Fellow at Northwestern and Stanford Universities, and he is the author of numerous essays and books on history and politics.

JON L. KYL WAS elected in 1994 to the U.S. Senate, representing Arizona. He is a member of the Judiciary, the Energy and National Resources, and the Intelligence Committees. Senator Kyl is the chairman of the Technology, Terrorism, and Government Information

Subcommittee and of the Water and Power Subcommittee. He is also the deputy Senate whip. Before being elected to the Senate, Senator Kyl served four terms in the U.S. House of Representatives.

MARK P. LAGON is a senior foreign and defense policy analyst with the Republican Policy Committee of the House of Representatives. He is also an adjunct professor of national security studies at Georgetown University. Mr. Lagon worked as a research assistant to Jeane Kirkpatrick at the American Enterprise Institute and as a visiting government professor at Georgetown University. He is the author of *The Reagan Doctrine: Sources of American Conduct in the Cold War's Last Chapter*, published in 1994.

MARGARITA MATHIOPOULOS is the senior vice president and head of the Corporate Communication and International Relations Division at Norddeutsche Landesbank in Hannover. She was a professor at the Universities of Braunschweig and Hannover, the Free University of Berlin, and Humboldt University. Ms. Mathiopoulos is the author of several books, including *On the Fragility of Democracy* and *Rendezvous with GDR: Political Myths and Their Demythologization.* She is the vice president of the German-Atlantic Association in Bonn.

ROBERT MROZIEWICZ is the under secretary of state at the Ministry of Foreign Affairs in Poland. He was the president of the United Nations Social and Economic Council and chairman of the First Committee of the United Nations General Assembly. Mr. Mroziewicz also served as the minister counselor and deputy permanent representative to the United Nations in 1990 and has been Poland's ambassador and permanent representative to the United Nations since 1991. He has been a professor of history and has written numerous articles.

WILLIAM E. ODOM is director of national security studies for the Hudson Institute and an adjunct professor at Yale University. As director of the National Security Agency from 1985 to 1988, he was responsible for the nation's signal intelligence and communications security. From 1981 to 1985, Mr. Odom served as deputy assistant and then assistant chief of staff for intelligence. From 1977 to 1981, he was the military assistant to the president's assistant for national security affairs, Zbigniew Brzezinski. On the National Security Council staff, Mr. Odom worked on strategic planning, Soviet affairs, nuclear weapons policy, telecommunications policy, and Persian Gulf security issues.

JOHN O'SULLIVAN is the founder and cochairman of the New Atlantic Initiative. He is editor at large of *National Review*, having been the editor of that magazine from 1988 to 1997. Mr. O'Sullivan was a special adviser to Prime Minister Margaret Thatcher, associate editor of the *Times* in London, assistant editor of the *Daily Telegraph* in London, and editor of *Policy Review*. He was the director of studies at the Heritage Foundation and a fellow at the Institute of Politics at Harvard University. Mr. O'Sullivan was made a Commander of the British Empire in 1991.

VALOIS V. PAVLOVSKIS is the president of the Baltic American Freedom League. In 1992 he was the deputy minister of defense in the newly independent Republic of Latvia, and in 1993 he was elected to Latvia's Parliament. Mr. Pavlovskis was the minister of defense from 1993 to 1994, and he also served on the Parliament's Defense and Internal Affairs Committee, the cabinet's Foreign Affairs and Defense Committee, and the Latvian National Security Council. He was the chairman of the American Latvian Association and vice president of the World Federation of Free Latvians.

RICHARD N. PERLE is a resident fellow at the American Enterprise Institute. He was the assistant secretary of defense for international security policy and the chairman of the North Atlantic Treaty Organization High Level Defense Group from 1981 to 1987. His recent work has concerned the future of NATO, the Bosnian war, and U.S. foreign policy. Mr. Perle is codirector of the AEI Commission on Future Defenses, a group organized to explore the use of advanced technology to increase the productivity of the armed forces. He is a contributing editor of *U.S. News & World Report* and a consultant to the secretary of defense. Mr. Perle is the editor of *Reshaping Western Security* (1991) and the author of *Hard Line* (1992), a political novel.

CHARLES POWELL was the foreign affairs and defense adviser to British Prime Minister Margaret Thatcher from 1983 to 1990 and to Prime Minister John Major until 1991. He is a director of several major companies, including the Jardine Matheson Holdings, National

Westminster Bank, Arjo Wiggins Appleton, and Möet Hennessy Louis Vuitton.

DAN QUAYLE was vice president of the United States during the Bush administration. He is a former member of the U.S. Senate and the U.S. Congress. He is the author of *Standing Firm*, his bestselling, vice-presidential memoir, and *The American Family: Discovering the Values That Make Us Strong*. Mr. Quayle also writes a nationally syndicated newspaper colum.

PETER W. RODMAN is director of National Security Programs at the Nixon Center for Peace and Freedom in Washington, D.C., and senior editor of *National Review*. He has served in the U.S. government, including as deputy assistant to the president for national security affairs and as director of the State Department's policy planning staff. Mr. Rodman was a special assistant to Henry Kissinger in the 1970s. He is the author of *More Precious than Peace: The Cold War and the Struggle for the Third World* and *Broken Triangle: Russia, China, and America after 25 Years*.

JEREMY D. ROSNER is special adviser to the president and secretary of state for NATO Enlargement Ratification. He was senior associate at the Carnegie Endowment for International Peace in Washington, and before that he was special assistant to President Clinton, in which position he served as counselor and senior director for legislative affairs on the staff of the National Security Council. Mr. Rosner was also President Clinton's principal speechwriter on national security topics. He is the author of *The New Tug-of-War: Congress, the Executive Branch, and National Security* and numerous articles.

RAINER SCHUWIRTH is deputy chief of staff of Germany's armed forces staff. He is also defense adviser to the German Permanent Delegation to NATO in Brussels. Major General Schuwirth enlisted in Germany's federal armed forces in 1964 and has held a number of positions including commander, branch chief, and military assistant to the federal minister of defense.

MARTIN SIEFF, senior State Department correspondent for the *Washington Times*, was twice nominated for the Pulitzer Prize for international reporting. He covered the collapse of communism as chief

Washington Times Soviet and East European correspondent from 1986 to 1992. He is also Washington correspondent for the *London Jewish Chronicle.*

RADEK SIKORSKI is an adviser to Jan Olszewski, the former Polish prime minister, and is the leader of the Movement for the Reconstruction of Poland. He was deputy minister of defense in Poland's first fully democratically elected government since World War II. Mr. Sikorski is the roving correspondent for the *National Review*, and his articles have appeared in the *Wall Street Journal*, *Foreign Affairs*, and *Rzeczpospolita*, the Polish newspaper of record. He is also the author of *Full Circle: A Homecoming to Free Poland.*

KALEV STOICESCU is Estonian ambassador to the United States. He began his career in the Ministry of Foreign Affairs of the Republic of Estonia in 1991 and has held a number of positions, which include the director of policy planning division of MFA, Tallin; the ambassador to the CSCE, Vienna; counselor and head of mission to the CSCE, Vienna; counselor on CSCE affairs, Department of International Organizations of MFA, Tallin; and the first secretary, Department of Information of MFA, Tallin.

NORMAN STONE is professor of modern history at the University of Oxford and a professor of history in the Department of International Relations at Bilkent University, Ankara. He was previously a fellow at Worcester College, Oxford. He was a fellow at Trinity College, Cambridge, and a fellow and director of studies in history at Jesus College, Cambridge. He is the author of several books, including *The Other Russia* (with Misha Glenny) and *Czechoslovakia: Crossroads and Crises.*

RÉKA A. SZEMERKÉNYI is a research fellow at the Hungarian Institute of International Affairs in Budapest. She was a consultant at the Economic Development Institute of the World Bank in Washington, D.C., and an assistant at the Hungarian Mission to the United Nations in New York. From 1991 to 1993, Ms. Szemerkényi was a senior assistant in international relations to the state secretary of defense in the Ministry of Foreign Affairs. Her recent publications include "NATO in Transition" in *NATO Studies and Documents* and "Central European Civil-Military Reforms at Risk."

MARGARET THATCHER was prime minister of the United Kingdom from 1979 to 1990. She was a Conservative member of Parliament, representing Finchley, from 1959 to 1992, and was opposition leader from 1975 to 1979. Lady Thatcher was awarded the Order of Merit in 1990. She is the author of two volumes of memoirs: *The Downing Street Years* (1993) and *The Path to Power* (1995).

K. ERIK TYGESEN is ambassador of the Kingdom of Denmark to the United States. Mr. Tygesen was ambassador of the Kingdom of Denmark to the Federal Republic of Germany from 1989 to 1995 and head of the Danish Military Mission to the Allied Control Authorities at Berlin from 1989 to 1990. He has headed numerous Danish delegations to international negotiations, including the Uruguay Round and the Lomé negotiations.

PAUL WOLFOWITZ is dean of the Paul H. Nitze School of Advanced International Studies, at Johns Hopkins University. He has held a number of positions in the U.S. government, including under secretary of defense for policy under Secretary of Defense Dick Cheney, U.S. ambassador to the Republic of Indonesia, and assistant secretary of state for East Asian and Pacific Affairs at the Department of State.

MICHAEL ZANTOVSKY is a member of the Czech Senate and chairman of the ODA Party. He was previously Czech ambassador to the United States. Mr. Zantovsky was President Havel's director of policy and his press secretary. He was a founding member of Civic Forum, an umbrella movement that coordinated the overthrow of the Communist regime at the end of 1989, and a correspondent for Reuters.

The American Enterprise Institute for Public Policy Research

Founded in 1943, AEI is a nonpartisan, nonprofit, research and educational organization based in Washington, D. C. The Institute sponsors research, conducts seminars and conferences, and publishes books and periodicals.

AEI's research is carried out under three major programs: Economic Policy Studies; Foreign Policy and Defense Studies; and Social and Political Studies. The resident scholars and fellows listed in these pages are part of a network that also includes ninety adjunct scholars at leading universities throughout the United States and in several foreign countries.

The views expressed in AEI publications are those of the authors and do not necessarily reflect the views of the staff, advisory panels, officers, or trustees.

A Note on the Book

This book was edited by
Dana Lane and the publications staff
of the American Enterprise Institute.
The text was set in Palatino, a typeface
designed by the twentieth-century Swiss designer
Hermann Zapf. Alice Anne English set the type,
and Edwards Brothers, Incorporated,
of Lillington, North Carolina,
printed and bound the book,
using permanent acid-free paper.

The AEI Press is the publisher for the American Enterprise Institute for Public Policy Research, 1150 Seventeenth Street, N.W., Washington, D.C. 20036; *Christopher DeMuth*, publisher; *Dana Lane*, director; *Ann Petty*, editor; *Leigh Tripoli*, editor; *Cheryl Weissman*, editor; *Alice Anne English*, production manager.